Two Manuals Combined into One Book

Medical Evacuation

And

Combat Risk Management

Department of Defense

Pentagon
Publishing

***FM 4-02.2**

Field Manual
No. 4-02.2

Headquarters
Department of the Army
Washington, DC, 8 May 2007

MEDICAL EVACUATION

Contents

***This publication supersedes Chapters 1–7 and Appendixes A, B, D–F, K, L, and N of FM 8-10-6 dated 14 April 2000, and FM 8-10-26 dated 16 February 1999.**

Figures

Tables

Preface

This field manual (FM) provides doctrine, as well as techniques and procedures for conducting medical evacuation and medical regulating operations. Medical evacuation encompasses both the evacuation of Soldiers from the point of injury (POI) or wounding to a medical treatment facility (MTF) staffed and equipped to provide essential care in theater and further evacuation from the theater to provide definitive, rehabilitative, and convalescent care in the continental United States (CONUS) and the movement of patients between MTFs or to staging facilities. Medical evacuation entails the provision of en route medical care; supports the joint health service support (JHSS) system; and links the continuum of care. In addition, it discusses the difference between medical evacuation and casualty evacuation (CASEVAC), as well as coordination requirements for and the use of nonmedical transportation assets to accomplish the CASEVAC mission. This publication is intended for use by medical commanders and their staffs, command surgeons, and nonmedical commanders involved in medical evacuation operations.

Users of this publication are encouraged to submit comments and recommendations to improve this publication. Comments should include the page, paragraph, and line(s) of the text where the change is recommended. The proponent for this publication is the United States (US) Army Medical Department (AMEDD) Center and School (USAMEDDC&S). Comments and recommendations should be forwarded, in letter format, directly to **Commander, USAMEDDC&S, ATTN: MCCS-FCD-L, 1400 East Grayson Street, Fort Sam Houston, Texas 78234-5052** or by using the e-mail address: medicaldoctrine@amedd.army.mil.

This publication applies to the Active Army, the Army National Guard (ARNG)/Army National Guard of the United States (ARNGUS), and the U.S. Army Reserve (USAR), unless otherwise stated.

Unless this publication states otherwise, masculine nouns and pronouns do not refer exclusively to men.

Use of trade or brand names in this publication is for illustrative purposes only and does not imply endorsement by the Department of Defense (DOD).

The staffing and organization structure presented in this publication reflects those established in base tables of organization and equipment (TOE) and are current as of the publication print date. However, such staffing is subject to change to comply with manpower requirements criteria outlined in Army Regulation (AR) 71-32. Those requirements criteria are also subject to change if the modification table of organization and equipment (MTOE) is significantly altered.

This publication implements the following North Atlantic Treaty Organization (NATO) International Standardization Agreements (STANAGs) and American, British, Canadian, and Australian (ABCA) Quadripartite Standardization Agreements (QSTAGs):

TITLE	STANAG	QSTAG
Stretchers, Bearing Brackets, and Attachment Supports	2040	
Medical Employment of Air Transport in the Forward Area	2087	
Medical and Dental Supply Procedures	2128	
Minimum Labeling Requirements for Medical Materiel		436
Documentation Relative to Medical Evacuation, Treatment, and Cause of Death of Patients	2132	470
Road Movements and Movement Control	2454	

TITLE	STANAG	QSTAG
Orders for the Camouflage of the Red Cross and Red Crescent on Land in Tactical Operations	2931	
Aeromedical Evacuation	3204	

When amendment, revision, or cancellation of this publication is proposed which will affect or violate the international agreements concerned, the preparing agency will take appropriate reconciliatory action through international standardization channels. These agreements are available on request from the Standardization Documents Order Desk, 700 Robins Avenue, Building 4, Section D, Philadelphia, Pennsylvania 19111-5094.

The AMEDD is in a transitional phase with terminology. This publication uses the most current terminology; however, other FM 4-02-series and FM 8-series may use the older terminology. Changes in terminology are a result of adopting the terminology currently used in the joint and/or NATO and ABCA Armies publication arenas. Also, the following terms are synonymous and the current terms are listed first:

- Medical logistics (MEDLOG); health service logistics (HSL); and combat health logistics (CHL).
- Roles of care, echelons of care, and level of care.
- Combat and operational stress control (COSC) and combat stress control (CSC).
- Behavioral health (BH) and mental health (MH).
- Chemical, biological, radiological and nuclear (CBRN) and nuclear, biological, and chemical (NBC).
- Stability operations; stability, security, transition, and reconstruction (SSTR) operations; and stability operations and support operations.

Introduction

The Army Health System (AHS) is a complex system of interrelated and interdependent systems which provides a continuum of medical treatment from the POI or wounding through successive roles of health care to definitive, rehabilitative, and convalescent care in the CONUS, as required. *Medical evacuation* is the system which provides the vital linkage between the roles of care necessary to sustain the patient during transport. This is accomplished by providing en route medical care and emergency medical intervention, if required, and to enhance the individual's prognosis and to reduce long-term disability.

Medical evacuation occurs at the tactical, operational, and strategic levels and requires the synchronization and integration of service component medical evacuation resources and procedures with the DOD worldwide evacuation system operated by the United States Transportation Command (USTRANSCOM).

Army medical evacuation is a multifaceted mission accomplished by a combination of dedicated ground and air evacuation platforms synchronized to provide direct support (DS), general support (GS), and area support within the joint operations area (JOA). At the tactical level, organic or DS medical evacuation resources locate, acquire, treat, and evacuate Soldiers from the POI or wounding to an appropriate MTF where they are stabilized, prioritized, and, if required, prepared for further evacuation to an MTF capable of providing required essential care within the JOA.

Although the most recognized mission of Army medical evacuation assets is the evacuation and provision of en route medical care to battlefield wounded, the essential and vital functions of medical evacuation resources encompass many additional missions and tasks that support the JHSS system. Medical evacuation resources are used to transfer patients between MTFs within the JOA and from MTFs to United States Air Force (USAF) mobile aeromedical staging facilities (MASFs) or aeromedical staging facilities (ASFs); emergency movement of Class VIII, blood and blood products, medical personnel and equipment; and serve as messengers in medical channels.

Medical regulating provides the interface with the DOD worldwide medical evacuation system by determining the patient's destination (the MTF best suited to provide the required care) and scheduling the means to transport the patient with the required en route medical care. Formal medical regulating begins at Role 3, however technological advances in information management (IM)/information technology (IT) are permitting this capability to be used at Role 1 and Role 2 MTFs in some situations.

Chapter 1

Overview of Army Health System Operations and Medical Evacuation

The AHS is the Army component of the military health system (MHS). Its capabilities are focused on delivering health care across the continuum of military operations—from the POI or wounding, through the JOA, to the CONUS support base. The two missions of the AHS are to provide health service support (HSS) (casualty care, medical evacuation, and medical logistics) and force health protection (FHP) (casualty prevention). The AHS is focused on promoting wellness, preventing casualties due to disease and nonbattle injuries (DNBIs), and providing timely and effective casualty care and management. Medical evacuation is the key factor in ensuring the continuity of care provided to our Soldiers by providing en route medical care during evacuation, facilitating the transfer of patients between MTFs to receive the appropriate specialty care, and ensuring that scarce medical resources (personnel, equipment, and supplies [to include blood]) can be rapidly transported to areas of critical need on the battlefield.

SECTION I — ARMY HEALTH SYSTEM

1-1. The provision of AHS is governed by well established and time-tested principles and rules which ensure the care provided to our Soldiers is timely and effective. For an in-depth discussion of these principles, rules, and roles of medical care, refer to FM 4-02.

1-2. The AHS is comprised of 10 medical functions. They are—

- Medical command, control, communications, computers, and intelligence.
- Medical treatment.
- Medical evacuation.
- Hospitalization.
- Dental services.
- Preventive medicine (PVNTMED) services.
- Combat and operational stress control (COSC).
- Veterinary services.
- Medical logistics.
- Medical laboratory support.

1-3. Army health system resources are arrayed across the battlefield in successive levels of support. These successive levels have increased medical care capabilities at each higher level. Medical evacuation and the provision of en route medical care ensures an uninterrupted continuum of care is maintained while Soldiers are moved through the roles of medical care to the MTF best suited to treat the patient's specific injuries.

PRINCIPLES OF ARMY HEALTH SYSTEM

1-4. The principles of AHS provide a framework in which medical planners can ensure that comprehensive plans are developed to support the tactical commander's operation plan (OPLAN). The principles of AHS are conformity, continuity, control, proximity, flexibility, and mobility. For this publication, the discussion of these principles has been focused toward the medical evacuation mission. For a general discussion of how these principles relate to the overall AHS mission, refer to FM 4-02.

CONFORMITY

1-5. Participating in the development of the OPLAN or the operation order (OPORD) ensures that the medical planner conforms to strategic, operational, and tactical plans. This is the most fundamental element for effectively providing AHS and ensures medical influence over the execution of medical evacuation operations. Only by participating in the orders process and developing a medical evacuation plan, will the medical planner ensure that medical evacuation support is arrayed on the battlefield in the right place at the right time and synchronized across operational commands to maximize responsiveness and effectiveness.

CONTINUITY

1-6. En route medical care provided during medical evacuation must be effective and continuous to prevent interruptions in the continuum of care. An interruption in medical treatment may result in an increase in morbidity, mortality, and disability. No patient is evacuated any farther than his physical condition or the military situation requires.

1-7. Medical evacuation resources provide the linkage between the roles of care within the JOA. They also provide interface with other deployed elements of the MHS operated by other services to enhance and facilitate the continuum of care from the POI to the CONUS support base.

CONTROL

1-8. Medical planners must ensure medical control is exercised over the execution of ground medical evacuation operations and that medical influence (technical and operational supervision) is exercised over the execution of aeromedical evacuation (AE) operations. Furthermore, medical planners must ensure the medical evacuation system is responsive to changing requirements and tailored to effectively support the forces within an assigned area of operations (AO). Since medical evacuation resources are limited, it is essential that medical control and influence be retained at the highest level consistent with the tactical situation.

1-9. A thorough and comprehensive medical evacuation plan is essential to establishing and maintaining control of medical evacuation operations characterized by decentralized execution of the plan. The medical evacuation plan complies with the combatant command guidance and intent and maximizes the use of scarce medical evacuation resources. When directed by the combatant command, Army air and ground ambulances may support operations conducted by other services, allied and coalition partners, and the host nation (HN).

PROXIMITY

1-10. The location of medical evacuation assets in support of combat operations is dictated by orders and the tactical situation (mission, enemy, terrain and weather, troops and support available, time available, and civil considerations [METT-TC]). Accurately determining time and distance factors and the availability of evacuation resources are critical to determining the disposition of evacuation assets. The speed with which medical evacuation is initiated is extremely important in reducing morbidity, mortality and disability. Medical evacuation time must be minimized by the effective disposition of resources, ensuring close proximity of both supported elements and MTF. Medical evacuation assets cannot be located so far forward that they interfere with the conduct of combat operations. Conversely, they must not be located so far to the rear that medical treatment is delayed due to lengthy evacuation routes.

1-11. Medical evacuation resources, both ground and air, are arrayed on the battlefield to best support both the tactical commander and the AHS. Depending upon the situation, evacuation resources may be placed in a DS role to support maneuvering forces or GS, during stability operations which are centrally located to accomplish an area support mission.

FLEXIBILITY

1-12. Changes in tactical plans or operations may require redistribution or reallocation of medical evacuation resources. Therefore, the medical evacuation plan must be designed to ensure flexibility and agility as well as enhance the ability to rapidly task-organize and relocate medical evacuation assets to meet changing battlefield requirements. Medical planners must also ensure medical control and influence is exercised through the orders process and facilitates the synchronization of air and ground evacuation assets to rapidly clear the battlefield.

MOBILITY

1-13. Medical evacuation assets must have the same mobility and survivability capability (such as armor protection) as the forces supported. This mobility and survivability ensures that medical evacuation resources can rapidly respond and that evacuation routes do not become too lengthy. Medical evacuation assets also enhance the mobility of forward deployed MTFs by rapidly evacuating their patients to the next role of medical care.

BATTLEFIELD RULES

1-14. The AMEDD has developed a set of battlefield rules to aid in establishing priorities and to resolve competing priorities within AHS activities. These rules are intended to guide the medical planner to resolve system conflicts encountered in designing and coordinating AHS operations. Although medical personnel seek always to provide the full scope of AHS services and support in the best possible manner, during every combat operation there are inherent possibilities of conflicting support requirements. The planner or operator applies these rules to ensure that the conflicts are resolved appropriately. These battlefield rules are depicted in Table 1-1.

Table 1-1. Army Medical Department Battlefield Rules

AMEDD BATTLEFIELD RULES
BE THERE (MAINTAIN A MEDICAL PRESENCE WITH THE SOLDIER)
MAINTAIN THE HEALTH OF THE COMMAND
SAVE LIVES
CLEAR THE BATTLEFIELD OF CASUALTIES
PROVIDE STATE-OF-THE-ART MEDICAL CARE
ENSURE EARLY RETURN TO DUTY OF THE SOLDIER

BE THERE

1-15. Ensure that medical evacuation assets are in close proximity to supported elements to enhance response time, increase Soldier confidence and be a combat multiplier. This is accomplished by complementing organic medical evacuation assets with medical evacuation assets placed in DS, GS, and area support roles.

Maintain the Health of the Command

1-16. Ensure that medical evacuation assets are in close proximity to the MTF. The closer evacuation assets are to an MTF, the more rapidly care may be rendered.

Save Lives

1-17. Ensure medical evacuation assets are the primary means of evacuating patients. Medical evacuation assets provide en route medical care that provides a continuum of care that is instrumental in preserving life and reducing long-term disability. The use of CASEVAC should only be used in extreme emergencies or when the medical evacuation system is overwhelmed. Refer to paragraphs 1-27 through 1-33 for a discussion on the differences between medical evacuation and CASEVAC operations.

Provide State-of-the Art Medical Care

1-18. Ensure sufficient medical evacuation assets are available to facilitate the movement of patients between MTFs (intratheater evacuation) and the emergency movement of Class VIII, blood and blood products, and medical personnel and equipment. This capability permits the medical planner to maximize the use of scarce medical resources and exercise economy of force without sacrificing state-of-the-art medical care.

Early Return To Duty of The Soldier

1-19. Ensure seamless integration with the intertheater medical evacuation system. This allows rapid evacuation and continuous medical treatment throughout the continuum of care resulting rapid return to duty (RTD) rates.

SECTION II — MEDICAL EVACUATION

PURPOSE

1-20. An efficient and effective medical evacuation system—

- Minimizes mortality by rapidly and efficiently moving the sick, injured, and wounded to an MTF.
- Serves as a force multiplier as it clears the battlefield enabling the tactical commander to continue his mission with all available combat assets.
- Builds the morale of Soldiers by demonstrating that care is quickly available if they are wounded.
- Provides en route medical care that is essential in improving the prognosis and reducing disability of the wounded, injured, or ill Soldiers.
- Provides medical economy of force.
- Provides connectivity of the AHS as appropriate to the MHS.

Tactical	Operational	Strategic
Corps Division BCT Emergency Class VIII Resupply Emergency Movement of Medical Personnel/Equipment Stability Operations	Coalition Joint Class VIII Resupply Joint Blood Program DOD Civilians/Contractors Medical Personnel/Equipment Military Working Dogs Homeland Security Operations Shore-to-Ship Stability Operations	Host Nation Nongovernment Organizations Private Volunteer Organizations DOD support to Stability Operations International Disaster Relief

Figure 1-1. Army medical evacuation as it supports full spectrum operations

1-21. The AHS is established in roles of increasing capability from the POI to definitive care. On the battlefield, casualties are evacuated rearward from one role of care to the next, the sequencing of this movement is dependent upon METT-TC factors. In a contiguous battlefield, well established lines of communication (LOC), large numbers of casualties, and wide array of MTFs can result in very deliberate evacuation from one sequential role to the next higher. However in many situations, such as noncontiguous battlefield, the array of medical resources across the battlefield, the availability of medical evacuation resources, and the number of patients being evacuated may facilitate procedures that permit bypassing roles of care in order to ensure the timely treatment and care of casualties. This evacuation plan will be established by the appropriate level of command in coordination with the command surgeon to ensure the best treatment is provided to all casualties.

ATTRIBUTES

DEDICATED RESOURCES

1-22. The Army medical evacuation system is comprised of dedicated air and ground evacuation platforms. These platforms have been designed, staffed, and equipped to provide en route medical care to patients being evacuated and are used exclusively to support the medical mission. The focus of the medical evacuation mission coupled with the dedicated platforms permit a rapid response to calls for support. The dedicated nature of this mission dictates that Army medical evacuation unit's posture themselves in a "ready alert" status, ready to rapidly respond to evacuation missions and not diverted to perform any other task. Medical evacuation resources are protected under the provisions of the Geneva Conventions from intentional attack by the enemy. (For a discussion of the Geneva Conventions in relation to air and ground evacuation operations, refer to Appendix A.)

EN ROUTE MEDICAL CARE

1-23. En route care is provided on all Army medical evacuation platforms. This care is essential for minimizing mortality, enhancing survival rates, and reducing disability of wounded, injured, or ill Soldiers. Refer to paragraphs 1-27 through 1-33 for a discussion on the differences between medical evacuation and CASEVAC operations.

CONNECTIVITY

1-24. The Army provides connectivity to the MHS. The synchronized employment of medical evacuation resources provides and maintains the seamless continuum of care from the POI through successive roles of essential care within the theater. In addition to evacuating patients and providing en route medical care, medical evacuation resources provide for the emergency movement of scarce medical resources such as critical Class VIII, blood, medical personnel, and medical equipment. Further, medical evacuation resources are used to transfer patients from one MTF to another within the theater, to facilitate specialty care as well as transferring patients from an MTF to a MASF to facilitate intertheater evacuation. Figure 1-1 depicts the connectivity provided by medical evacuation platforms between the roles of care within theater. In addition to providing connectivity within the AHS, Army medical evacuation resources provide medical evacuation support and interface with MTFs of the other services deployed in the theater. In joint and multinational operations, the geographic combatant commander (GCC) may direct Army assets to provide this support and connectivity to joint, allied, and coalition forces within the JOA.

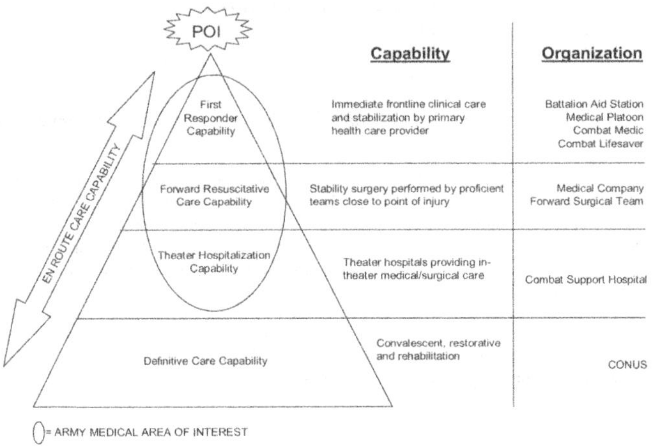

Figure 1-2. Army air and ground evacuation platforms provide connectivity to assure a seamless continuum of medical care

MEDICAL ECONOMY OF FORCE

1-25. The flexibility and versatility which medical evacuation resources afford the AHS to respond to urgent requirements throughout the battlefield is essential in ensuring the provision of seamless medical care. Medical economy of force is achieved for low density, high demand medical specialties (such as a neurosurgeon), medical supplies, and medical equipment (such as computer tomography scans or magnetic resonance imaging equipment) by having the capability to move the patient to the required care over long distances and by permitting the emergency of cross-leveling on medical supplies reducing the need for large Class VIII stockpiles. By providing medical economy of force, the deployed medical footprint is reduced without negatively impacting on the care provided to the Soldier.

FORCE MULTIPLIER

1-26. A highly effective, efficient, and responsive AHS increases Soldiers' confidence that if wounded in battle, they will receive timely and appropriate medical care to enhance their prognosis, speed their recovery, and reduce long-term disabilities. Medical evacuation resources locate and acquire wounded Soldiers and facilitate their entry into the AHS. These resources clear the battlefield of casualties which enables the tactical commander the ability to capitalize on and exploit battlefield opportunities with all available assets in the execution of his warfighting mission.

SECTION III — MEDICAL EVACUATION VERSUS CASUALTY EVACUATION

MEDICAL EVACUATION

1-27. Medical evacuation is performed by dedicated, standardized medical evacuation platforms, with medical professionals who provide the timely, efficient movement and en route care of the wounded, injured, or ill persons from the battlefield and/or other locations to MTFs. Medical evacuation is an AMEDD function that supports and is an integral part of the AHS. The provision of en route care on medically equipped vehicles or aircraft greatly enhances the patient's potential for recovery and may reduce long-term disability by maintaining the patient's medical condition in a more stable manner.

1-28. Medical evacuation ground/air ambulance platforms are defined as: Platforms especially for the medical evacuation mission with allocated medical equipment specifically designed for the purpose of enroute care and by trained medical personnel.

1-29. The gaining MTF in coordination with the losing MTF, is responsible for arranging for the evacuation of patients from the lower role of care. For example, Role 2 medical units are responsible for evacuating patients from Role 1 MTFs.

1-30. Medical evacuation begins when a request is initiated and continues until the patient is released from the AHS by medical authorities.

CASUALTY EVACUATION

1-31. Casualty evacuation is a term used to refer to the movement of casualties aboard nonmedical vehicles or aircraft.

WARNING

Casualties transported in this manner may not receive proper en route medical care or be transported to the appropriate MTF to address the patient's medical condition. If the casualty's medical condition deteriorates during transport, or the casualty is not transported to the appropriate MTF, an adverse impact on his prognosis and long term disability or death may result.

1-32. If dedicated medical evacuation platforms (ground and air) are available, casualties should be evacuated on these conveyances to ensure they receive proper en route medical care.

1-33. Since CASEVAC operations can reduce combat power and degrade the efficiency of the AHS, units should only use CASEVAC to move Soldiers with less severe injuries when medical evacuation assets are overwhelmed. Medical planners should ensure CASEVAC operations are addressed in the OPLAN/OPORD as a separate operation, as these operations require preplanning, coordination, synchronization, and rehearsals. The CASEVAC plan should ensure casualties with severe or life-threatening injuries are prioritized for evacuation and are evacuated on dedicated medical evacuation platforms.

1-34. When possible, nonmedical vehicles/aircraft transporting casualties should be augmented with a trauma specialist or combat lifesaver (CLS). (On nonmedical aircraft, sufficient space may not be available to permit a caregiver to accompany the casualties.) The type of en route monitoring and medical care/first aid provided is limited by the following factors:

- Skill level of the individual providing care. (The combat medic is military occupational specialty [MOS]-qualified [MOS 68W] to provide emergency medical treatment [EMT]; the CLS is trained to provide enhanced first aid.) The combat medic can provide emergency medical intervention, whereas the CLS can only monitor the casualty and ensure that the basic lifesaving first-aid tasks are accomplished.
- Medical equipment available.
- Number of casualties being transported.
- Accessibility of casualties—If nonstandard evacuation vehicle is loaded with the maximum number of casualties, the combat medic or CLS may not be able to attend to the casualties while the vehicle is moving. If the condition of a casualty deteriorates and emergency measures are required, the vehicle will have to be stopped to permit care to be given.

SECTION IV — THEATER EVACUATION POLICY

ESTABLISHING THE THEATER EVACUATION POLICY

1-35. The theater evacuation policy is established by the Secretary of Defense, with the advice of the Joint Chiefs of Staff and upon the recommendation of the theater commander. The policy establishes, in number of days, the maximum period of noneffectiveness (hospitalization and convalescence) that patients may be held within the theater for treatment. This policy does not mean that a patient is held in the theater for the entire period of noneffectiveness. A patient who is not expected to be ready for RTD within the number of days established in the theater evacuation policy is evacuated to the CONUS or other safe haven. This is done providing that the treating physician determines that such evacuation will not aggravate the patient's disabilities or medical condition. For example, a theater evacuation policy of seven days does not mean that a patient is held in the theater for six days and then evacuated. Instead, it means that a patient is evacuated as soon as possible after the determination is made that the Soldier cannot be returned to duty within seven days following admission to a Role 3 hospital.

TEMPORARY REDUCTIONS

1-36. To the degree that unplanned for increases in patients occur (due perhaps to an epidemic or heavy combat casualties), a temporary reduction in the policy may be necessary. This reduction is used to adjust the volume of patients being held in the theater hospital system. A reduction in the evacuation policy increases the number of patients requiring evacuation out-of-theater and increases the requirement for evacuation assets. This action is necessary to relieve the congestion caused by the patient increases. A decrease in the theater evacuation policy increases the evacuation asset requirements.

ESSENTIAL CARE IN THEATER AND LENGTH OF POLICY

1-37. Due to the reduction of the medical footprint within the theater, health care delivery to deployed forces is now provided under the concept of essential care in theater. This concept provides for essential and stabilizing care being rendered within the theater, with the patient being evacuated to CONUS or other safe haven for definitive, rehabilitative, and convalescent care. In current and ongoing operations, if the patient cannot be treated and returned to duty within seven days of admission to a combat support hospital (CSH), the patient is evacuated from the theater for continued care.

1-38. The time period established by the theater evacuation policy starts on the date the patient is admitted to the first hospital. The total time a patient is hospitalized in the theater (including transit time between MTFs) for a single, uninterrupted episode of illness or injury should not exceed the number of days stated in the theater evacuation policy. Though guided by the evacuation policy, the actual selection of a patient for evacuation is based on clinical judgment as to the patient's ability to tolerate and survive the movement to the next role of care.

Exception to Policy

1-39. An exception to the theater evacuation policy may be required with respect to SOF personnel. This exception may be required to retain low density MOS skills within the theater. Retaining these personnel within the theater for an extended period of time is possible if the medical resources are available within the theater to treat their injuries and provide for convalescence and rehabilitation. If retention within the theater would result in a deterioration of their medical condition or would adversely impact on their prognosis for full recovery, they are evacuated from the theater for definitive care.

Evacuation Policy Considerations

1-40. Paragraphs 1-41 through 1-45 discuss the various considerations of the theater evacuation policy.

Physicians and Dentists

1-41. To physicians and dentists engaged in direct patient treatment and decisions relating to patient disposition, it means that there is a maximum period which clinical staffs may complete the necessary treatment needed to return the patient to full duty within the theater. If the theater policy is seven days and full RTD can be predicted within that time, the patient is retained in the theater. If the patient cannot be returned to full duty within seven days, the patient is evacuated out-of-theater as early as clinically prudent.

Medical Planners and Staff

1-42. The medical staff computes mix, number, and distribution of hospital beds required in the theater.

1-43. The medical operator has a management tool which, when properly adjusted and used, provides the balance between patient care and tactical support requirements. The medical staff is able to tailor a medical package specifically designed to handle the patient workloads, with maximum benefit to the patients and with maximum economy of available resources.

Logisticians

1-44. The nonmedical logistician is able to estimate his total obligation to support this system.

United States Air Force Planner

1-45. The USAF planner can accurately plan the USAF AE requirements for both intra- and intertheater patient movements.

FACTORS DETERMINING THE THEATER EVACUATION POLICY

1-46. To fully understand how the theater evacuation policy affects AHS operations, the medical operator should be aware of the factors that influence the establishment of this policy.

Nature of the Operation

1-47. A major factor in determining the theater evacuation policy is the nature of the tactical operations. For example: Will the operations be of short duration and with a low potential for conflict? Will there be operations of long duration with significant combat operations? Will chemical, biological, radiological, nuclear, and high yield explosives (CBRNE) be employed? Will only conventional weapons be used? Is a static combat situation expected? Is there a significant threat of terrorist activities? Are the majority of patients anticipated to be DNBI patients or those with combat-related trauma?

Number and Type of Patients

1-48. Another factor affecting the policy is the number and types of patients anticipated and the rate of patient RTD. Admission rates vary widely in different geographical areas of the world and in different types of military operations.

Evacuation Means

1-49. The means (quantity and type of transportation) available for evacuation of patients from the theater to CONUS is an essential factor impacting on the evacuation policy.

Availability of Replacements

1-50. The capability of CONUS to furnish replacements to the theater is another consideration. For each patient who is evacuated from the theater to CONUS, a fully trained and equipped replacement must be provided. During a small-scale conflict overseas, the CONUS replacement capability is much greater than when compared to a large-scale conflict such as World War II.

Availability of In-Theater Resources

1-51. Limitations of all AHS resources (such as insufficient number and types of medical units in the division/corps to support the brigade combat teams [BCTs] and an insufficient amount of medical and nonmedical logistics) will have an impact on the theater evacuation policy. The more limitations (or shortages), the shorter the theater evacuation policy.

IMPACT OF THE EVACUATION POLICY ON ARMY HEALTH SYSTEM REQUIREMENTS

Shorter Evacuation Policy

1-52. A shorter theater evacuation policy—

- Results in fewer hospital beds required in the theater and a greater number of beds required elsewhere.
- Creates a greater demand for intertheater USAF evacuation resources. (A shortened intratheater evacuation policy would likewise increase the number of airframes required in the theater.)
- Increases the requirements for replacements to meet the rapid personnel turnover which could be expected, especially in combat units. (The impact this would have on both intra- and intertheater transportation and other requirements must also be considered.)

Longer Evacuation Policy

1-53. A longer theater evacuation policy—

- Results in a greater accumulation of patients and a demand for a larger AHS infrastructure in the theater. It decreases bed requirements elsewhere.
- Increases the requirements for medical logistics (medical supplies, equipment, and equipment maintenance) and nonmedical logistics support.
- Increases the requirements for hospitals, engineer support, and all aspects of base development for deployed AHS force. (It demands the establishment of a larger number of hospitals within the theater and may require medical specialty augmentation.)
- Provides for a greater proportion of patients to RTD within the theater and, thus, reduces the loss of experienced manpower.
- A longer intratheater evacuation policy may decrease the demand on the intratheater evacuation assets and system.

1-54. The concept of essential care in theater does not support longer evacuation policies as the deployed hospitals are not designed to provide definitive, rehabilitative, and convalescent care/services. If the theater evacuation policy is extended in theaters operating under the essential care in theater concept, augmentation of medical specialty resources will be required.

Patient Stabilization

1-55. The evacuation policy has no impact on the patient stabilization period for movement. This period is known as the *evacuation delay*. It is the period of time planned for between the time of patient reporting and the time of medical evacuation of the patient to the next role of care. Evacuation delays normally range from 24 to 72 hours and are designated by the Army service component command (ASCC) surgeon.

Chapter 2

Medical Evacuation Resources

This chapter discusses the mission, functions, and capabilities of medical evacuation units and elements, as specified in the unit's TOE. It also discusses the command and control (C2) headquarters to which they are assigned. The discussion of each organization also includes the basis of allocation (BOA), limitations, and dependencies.

SECTION I — MANEUVER BATTALION MEDICAL PLATOON AMBULANCES

MEDICAL PLATOON AMBULANCE SQUADS

2-1. The ground ambulance team is the basic medical evacuation element used within the BCTs and at the corps and theater AOs. These ground ambulance teams provide medical evacuation from the POI to supporting MTFs while ensuring the continuity of care en route. Ambulance squads consisting of two ambulance teams organic to the medical platoons of movement and maneuver and fires units and to the evacuation platoon of the brigade support medical companies (BSMCs).

2-2. The primary mission of the maneuver medical platoon ambulance squads is to provide ground ambulance evacuation support from supported infantry/armor companies or from POI back to a casualty collection point (CCP) or to the Role 1 MTF/battalion aid station (BAS). Maneuver medical platoon ambulance teams are assigned to heavy BCT (HBCT), infantry BCT, Stryker BCT (SBCT), reconnaissance squadrons, and reconnaissance and target acquisition (RSTA) squadrons. They also provide area support to other elements (which do not have organic medical evacuation resources) operating in their AO. Ground ambulance support provided is consistent with evacuation precedence, tactical standing operating procedures (TSOPs), and other operational considerations.

2-3. For definitive information on maneuver battalion medical platoon operations, see FM 4-02.4, FM 4-02.6, and FM 4-02.21.

MANEUVER UNIT MEDICAL PLATOON

ORGANIZATION

2-4. The medical platoon consists of a platoon headquarters section, a treatment squad, combat medic section, and an ambulance squad. The number of ambulance squads in an ambulance platoon varies and is based on the type of parent organization. The infantry, airborne, and air assault battalions' ambulance platoons have two ambulance squads equipped with high mobility multipurpose wheeled vehicle (HMMWV) ambulances. The mechanized infantry and armor combat maneuver battalions' ambulance platoons have four ambulance squads equipped with M-113 track ambulances. The SBCT infantry battalion's and RSTA squadron's ambulance platoons have two ambulance squads equipped with Stryker wheeled armored ambulances (referred to as the Stryker medical evacuation vehicle [MEV]).

Medical Platoon Headquarters

2-5. The medical platoon leader is a physician and also serves as the battalion surgeon. He is assisted by the medical operations officer (field medical assistant) in the operational, administrative, and logistical support aspects of the platoon.

Ambulance Section

2-6. The ambulance section of the medical platoon is organized into ambulance squads and is supervised by the platoon sergeant. Each squad contains a noncommissioned officer (NCO) squad leader, three combat medics, and two ambulances.

Ambulance Squad Leader

2-7. The ambulance squad leader ensures that ambulance team members perform operator maintenance on assigned vehicles. An ambulance team consists of one vehicle and two medical personnel (an aide/evacuation NCO and a combat medic).

Ambulance Team

2-8. Ambulance team members are also responsible for performing operator maintenance on assigned ambulances. Specific duties of the ambulance team are to—

- Administer EMT as required.
- Initiate the Department of Defense (DD) Form 1380, US Field Medical Card (FMC), as required.
- Prepare patients for ground and air medical evacuation.
- Provide medical evacuation of wounded or injured Soldiers from the POI to supporting MTF or ambulance exchange point (AXP).
- Provide patient care en route.
- Provide emergency movement of medical personnel and emergency delivery of blood, medical supplies, and medical equipment.
- Assist in the care of COSR casualties.
- Maintain operational readiness of the ambulance.
- Resupply company and platoon trauma specialists, when required.
- Operate the vehicle and maintain contact with supported and supporting elements.
- Find and collect the wounded.
- Serve as messengers within medical channels.

PRIMARY FUNCTIONS OF AMBULANCE TEAMS

2-9. The primary function of the ambulance team is to collect and treat the sick, injured, and wounded Soldiers on the battlefield and to provide medical evacuation support from the POI, CCP, or AXP to the supporting MTF.

SECTION II — AMBULANCE/EVACUATION PLATOON—FORWARD SUPPORT, BRIGADE SUPPORT, OR AREA SUPPORT MEDICAL COMPANY

MEDICAL COMPANIES

2-10. Medical companies provide Role 2 MTF and medical evacuation. Medical companies assigned to or in DS of a BCT are referred to as either a forward support medical company (FSMC) or a BSMC. Area support medical companies (ASMCs) are assigned to multifunctional medical battalions and operate Role 2 MTF and provide medical evacuation support for division, and theater units. In addition, all medical companies provide Role 1 area medical support to units without organic medical personnel and may augment/reinforce maneuver medical platoons/sections with treatment and medical evacuation support.

FORWARD SUPPORT MEDICAL COMPANY

2-11. The FSMC, an Army of Excellence (AOE) designed unit is organized to provide triage and management of mass casualties, advanced trauma management (ATM), initial resuscitation and stabilization, care for patients with DNBI, and battle wounded and injured Soldiers. The FSMC also provides intervention for combat and operational stress disorders to include COSR and preparation of patients for further medical evacuation. The FSMC is organized into a company headquarters, a treatment platoon, an ambulance platoon, a PVNTMED section, and a BH section.

2-12. The ambulance platoon of FSMC-Light and Airborne units employ a platoon headquarters with four-wheeled ambulance squads (eight ambulances); the FSMC-Air Assault employs a platoon headquarters with three-wheeled ambulance squads (six ambulances); the FSMC-Heavy employs a platoon headquarters with three-tracked ambulance squads (six ambulances) and two-wheeled ambulance squads (four ambulances); and the FSMC-Airborne employs a platoon headquarters with four-wheeled ambulance squads (eight ambulances). A complete discussion on the organization, mission, and functions of the FSMC is provided in FM 4-02.6.

BRIGADE SUPPORT MEDICAL COMPANY

2-13. The new BSMC is designed to support the modular BCT. The BOA for the BSMC is one per maneuver brigade supported. The mission of the BSMC is to provide HSS to all BCT units operating within the brigade AO. The BSMC locates and establishes its company headquarters and also operates a Role 2 MTF and may operate Role 1 MTFs on an area support basis for units that do not have organic medical assets. The company provides C2 for its organic and attached/operational control (OPCON) medical augmentation elements. The BSMC may be augmented with a forward surgical team (FST) providing the company surgical capability, see FM 4-02.25. The BSMC is organized into a company headquarters, a PVNTMED section, a BH section, a treatment platoon, and an evacuation platoon.

2-14. The evacuation platoon performs ground medical evacuation and en route patient care for the supported units. The evacuation platoon consists of a platoon headquarters, a GS evacuation section, and a DS evacuation section. The platoon employs five HMMWV evacuation squads (or ten evacuation teams). A complete discussion on the organization, mission, and functions of the BSMC is provided in FM 4-02.6 and Field Manual Interim (FMI) 4-90.1.

AREA SUPPORT MEDICAL COMPANY

2-15. The ASMC performs functions similar to those of the BCT medical companies. The ASMCs are employed primarily in support of division, corps, and echelons above corps (EAC) units. They are deployed to a geographical area to provide area HSS, or may be deployed to provide HSS for designated units. The ASMC also establishes its Role 2 MTF in a secure location centrally located for supported units. Medical treatment squads/teams of the ASMC may be deployed to establish Role 1 MTF and provide HSS support to concentrations of nondivisional units that do not have organic medical capabilities. The ASMC is organized into a company headquarters, a treatment and ambulance platoon, and a BH section.

2-16. The ambulance platoon performs ground medical evacuation and en route patient care for supported units. The ambulance platoon consists of a platoon headquarters, four ambulance squads (or eight ambulance teams), one HMMWV control vehicle, and eight HMMWV ambulances. A complete discussion on the organization, mission, and functions of the ASMC is provided in FM 4-02.24.

ORGANIZATIONS

2-17. The BSMC/FSMC normally establishes its Role 2 MTF in the BSA, while the ASMC normally establishes its Role 2 MTF in echelons above brigade (EAB) and echelons above division sustainment areas and at theater level. The medical company provides DS/GS ground ambulance evacuation for supported units. It coordinates/ requests air ambulance support for air evacuation of patients from supported Role 1 MTFs/BASs, POI, CCP, and AXPs. The organizational structure, number and type of ambulances assigned to an ambulance/evacuation platoon may differ depending on their location on the battlefield and the types of units supported. For example, a BSMC in support of a HBCT has a mix of armored and wheeled evacuation vehicles while a BSMC in support of the SBCT has only wheeled evacuation vehicles.

2-18. The company headquarters section is organized into a command element; a support element; a unit supply element; a medical supply and medical maintenance element; and an operations and communications element.

Treatment Platoon

2-19. The treatment platoon is composed of a platoon headquarters, treatment squads, and an area support section, which consists of an area support squad, an area support treatment squad, and an area patient holding squad.

2-20. The platoon headquarters is the C2 element of the platoon. The platoon headquarters:

- Determines and directs the disposition of patients and submits requests through the company command post (CP) for their evacuation of patients to supporting hospitals.
- Directs, coordinates, and supervises platoon operations.
- Directs the activities of the Role 2 MTF, monitors Class VIII supplies, blood usage and inventory levels, and keeps the commander informed.
- Manages platoon operations, operations security (OPSEC), communications, medical administration, organizational training, supply transportation, patient accountability, statistical reporting functions, and blood situation reporting.

2-21. The treatment squad element can contain up to four treatment squads, depending on the type company assigned. The treatment squads have the capabilities to:

- Provide emergency and routine sick call treatment services to Soldiers assigned to supported units.
- Operate as separate treatment teams (Teams A and B) for limited periods of time.
- Operate up to four treatment stations.
- Reinforce or reconstitute similar treatment squads and can operate for up to 48 hours while separated from their parent unit.

2-22. For medical companies deployed in the BSA, these squads can further be used to—

- Provide augmentation to maneuver battalion medical platoons.
- Facilitate the movement of the treatment platoon MTF from one location to another or echelons forward to establish an MTF at a new location. A treatment squad can be echeloned forward to establish an MTF at a new location. The echeloning of elements allows the old treatment site to remain operational until the new site is established.
- Augment an AXP, CCP, or positioned at designated points in the ambulance shuttle system, when required.

2-23. The area support section of the treatment platoon is composed of an area treatment squad, an area support squad, and a patient-holding squad. These squads form the medical company Role 2 MTF. The area support treatment squad provides trauma care and routine sick call services to personnel assigned to units located in the supported area of the medical company. The area support squad provides operational dental care that includes emergency and essential dental services, limited laboratory and radiological services, and blood support commensurate with Role 2 MTF. The patient-holding squad provides up to 40 patient beds (40 patient beds for heavy divisions and 20 patient beds for light divisions) for patients requiring minimal treatment. Patients held in the patient-holding area are those patients who are expected to be RTD within 72 hours from the time they are held for treatment.

2-24. The area treatment squad is identical in personnel and equipment as the treatment squads of the treatment section. It is the base medical treatment element of a Role 2 MTF. It provides sick call services and initial resuscitative treatment for supported units. For communications, the squad employs frequency modulated (FM) radios and is deployed in the medical radio and wires communications nets.

Ambulance/Evacuation Platoon

2-25. The ambulance/evacuation platoon provides C2 for ambulance platoon operations. The ambulance platoon headquarters element maintains communications to direct ground ambulance evacuation of patients. It provides ground ambulance evacuation support for supported maneuver battalions and for supported units operating in the sustainment areas. The ambulance headquarters element performs route reconnaissance and develops and issues graphic overlays to all its ambulance teams. It also coordinates and establishes AXPs for both air and ground ambulances, as required.

2-26. Ambulance squads provide ground ambulance evacuation of patients from supported BASs/unit aid stations back to the Role 2 MTF that is located in the brigade support area (BSA). An ambulance squad consists of two ambulance teams (two ambulances, wheel or tracked vehicles). Ambulance squad personnel—

- Perform EMT, evacuate patients, and provide for their continued care en route.
- Operate and maintain assigned communication and navigational equipment.
- Perform preventive maintenance checks and services (PMCS) on ambulances and associated equipment.
- Maintain supply levels for the ambulance medical equipment sets (MES).
- Ensure that appropriate property exchange of medical items (such as litters and blankets) is made at sending and receiving MTF.
- Prepare patients for ground and air medical evacuation.
- Provide medical evacuation of wounded or injured Soldiers from the POI to supporting MTF, CCP or AXP.
- Initiate the DD Form 1380, FMC, as required.
- Provide emergency movement of medical personnel and emergency delivery of blood, medical supplies, and medical equipment.
- Assist in the care of COSR casualties.
- Maintain operational readiness of the ambulance.
- Resupply company and platoon trauma specialists, when required.
- Operate the vehicle and maintain contact with supported and supporting elements.
- Find and collect the wounded.
- Serve as messengers within medical channels.

Preventive Medicine Section

2-27. Preventive medicine sections are organic to the divisions and the BSMC of the BCT. The PVNTMED section is also found in the multifunctional medical battalion (MMB) headquarters. The PVNTMED section assigned to medical companies has a primary responsibility for supervising the unit's PVNTMED program as described in AR 40-5. The section—

- Ensures measures are implemented to protect personnel against food, water, and arthropodborne diseases, as well as environmental injuries.
- Provides advice and consultation in the area of health threat assessment, FHP, environmental sanitation, epidemiology, sanitary engineering, and pest management.
- Assists the higher headquarters in determining requirements for medical intelligence assessments, particularly with respect to CBRN and disease prevalence.
- Coordinates with supporting veterinary teams for conducting and implementing food safety and quality assurance surveillance and assisting in foodborne and zoonotic disease surveillance and control.

Behavioral Health Section

2-28. Combat and operational stress is controlled through vigorous prevention, consultation, and restoration programs by the BSMC BH section. These programs are designed to maximize the RTD rate of

COSR Soldiers by providing rest/restoration within or near their unit areas. The BSMC BH section has a primary responsibility for assisting commanders in controlling combat stress and serves as a consultant to the commander, staff, and others involved with providing prevention and intervention services to unit Soldiers and their families. For additional information on the operations and functions of BSMCs, and ASMCs, refer to FMs 4-02.6 and 4-02.24.

SECTION III — MEDICAL COMPANY (GROUND AMBULANCE)

OPERATIONAL INFORMATION

2-29. The AHS is a continuum of increasing roles of care extending from the POI through the CONUS base. All sick, injured, and wounded patients must be evacuated from the battlefield in the shortest possible time to the MTFs that can provide the required treatment. Ground ambulances serve as one of the primary means of evacuating patients from the battlefield.

MISSION

2-30. The mission of the medical company, ground ambulance, TOE 08453A00, is to provide ground evacuation within the theater.

ASSIGNMENT

2-31. The medical company, ground ambulance is normally assigned or attached to MMB (TOE 08485G00) or a medical brigade (MEDBDE) for C2.

EMPLOYMENT

2-32. The medical company, ground ambulance, provides DS to BCTs and is employed in the division, corps, and EAC to provide area support. It is tactically located where it can best control its assets and execute its patient evacuation mission.

BASIS OF ALLOCATION

2-33. The BOA within the combat zone is one per division supported and one per two divisions supported within the theater.

CAPABILITIES

2-34. At Level I, the unit is capable of providing:

- A single-lift capability for evacuation of 96 litter patients or 192 ambulatory patients.
- Medical evacuation from BCT/division medical companies and ASMCs to supporting hospitals.
- Reinforcement of BCT/division medical company evacuation assets.
- Reinforcement of covering force and unassigned area operations.
- Movement of patients between hospitals and aeromedical staging squadrons/MASFs, railheads, or seaports in the division, corps, and EAC.
- Area evacuation support beyond the capabilities of the ASMC.
- Emergency movement of medical personnel and supplies.
- Preparation of patients for ground and air medical evacuation.
- Medical evacuation of wounded or injured Soldiers from the POI to supporting MTF, CCP, or AXP.
- Field feeding and vehicle refueling support when collocated with the MMB.

DEPENDENCY

2-35. This unit is dependent upon the appropriate elements of the division, corps, EAC, or ASCC for—

- Religious, financial management, legal, personnel, mortuary affairs, and administrative services.
- Viable communications systems for C2 and adequate road networks.
- Laundry, shower, and clothing repair.
- Generator, communication equipment, and communication security (COMSEC) equipment maintenance.
- Health service support to include hospitalization.

ORGANIZATION AND FUNCTIONS

2-36. This company is organized into a company headquarters section, two ambulance platoon headquarters, and two evacuation sections (Figure 2-1).

2-37. The company headquarters provides C2, communications, administration, field feeding, and logistical support (to include maintenance) for the subordinate ambulance platoons.

2-38. The ambulance platoon headquarters provides C2 for the subordinate ambulance squads.

2-39. The evacuation section operates ambulances and provides en route medical care for patients. Each evacuation section consists of 12 ambulances, each with a two-man crew. The members of the squad maintain the level of expendable Class VIII supplies in the ambulance MES by reconstituting supplies from medical companies or hospitals when they pick up or drop off patients. They are also responsible for performing operator maintenance on assigned vehicles.

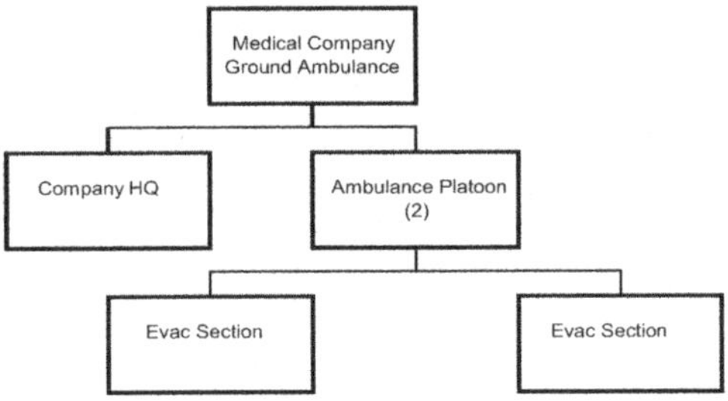

Figure 2-1. Medical company, ground ambulance

SECTION IV — MULTIFUNCTIONAL MEDICAL BATTALION

MODULAR FORCE

2-40. A modular force is a key aspect in the transformation of the Army. The modular force model is based on brigade-sized elements that are more responsive than division-sized elements and can perform joint and expeditionary-type missions. Using modular units, Army planners can tailor force structure, reduce strategic lift requirements, and create flexible forces with specialized capabilities based on everchanging mission requirements. To facilitate the synchronization and provision of AHS services and support, the AMEDD C2 organizations have been redesigned into a modular design. The MMB will replace the functional battalions (area support, evacuation, and medical logistics).

2-41. The new modular design of medical units gives the medical command (MEDCOM) the ability to—

- Assist in the employment of the optimal mix of medical capabilities within an expeditionary to campaign framework.
- Ensure seamless, state-of-the-art care, regardless of delivery location.
- Provide tested and proven systems to the battlefield and ensure the provision of the right care at the right place and time.
- Promote scalability through easily tailored, capabilities-based packages that result in improved tactical mobility, reduced footprint, and increased modularity for flexible task organization.
- Provide and enable the joint force commander (JFC) to choose among augmentation packages, thus enabling rapid synchronization of desired medical capabilities.

MULTIFUNCTIONAL MEDICAL BATTALION

MISSION

2-42. The mission of the MMB, TOE 08486G000, is to provide a scalable, flexible and modular C2, administrative assistance, logistical support, and technical supervision capability for assigned and attached medical organizations (companies, detachments, and teams) task-organized for support of deployed forces.

ASSIGNMENT

2-43. This unit is assigned to a MEDCOM or a MEDBDE.

EMPLOYMENT

2-44. The MMB is normally employed in a MEDCOM or MEDBDE AO.

BASIS OF ALLOCATION

2-45. The BOA is one per three to six subordinate company-size units plus the blood detachment.

CAPABILITIES

2-46. At Level I, this unit provides—

- Command and control, staff planning, supervision of operations, medical and general logistics support as required, and administration of the assigned and attached units conducting AHS operations in its supported AO.
- Task organization of medical assets to meet the projected patient workload.
- Advice to senior commanders in the AO on the medical aspects of their operations.
- Coordination of medical regulating and patient movement within the AO.
- Monitoring, planning, and coordinating ground and air evacuation within the battalion AO. Coordinating air evacuation support requirements and synchronizating the air evacuation plan into the overall medical evacuation plan with the supporting aviation unit.
- Consultation and technical advice on PVNTMED (medical entomology; medical, and occupational and environmental health [OEH] surveillance; and sanitary engineering), pharmacy procedures, COSC and BH, medical records administration, veterinary services, nursing practices and procedures, dental services, medical laboratory procedures, and automated medical information systems to supported units.
- Guidance for facility site selection and area preparation, as required.
- Monitoring and supervising medical logistics operations, to include Class VIII supply/resupply, medical equipment maintenance and repair support, optical fabrication and repair support, and blood management.
- Planning and coordination for Role 1 and Role 2 AHS operations, to include staff advice to units without organic medical assets within the AO.

- Unit level maintenance for wheeled vehicles and power generation equipment and wheeled vehicle recovery operations support to assigned or attached units.
- Organizational communications equipment maintenance support for the battalion.
- Food service support for staff and other assigned/attached medical units.
- Consolidated property book maintenance.
- Religious support for MMB staff, assigned/attached medical units, and casualties in MMB subordinate unit MTFs.

DEPENDENCY

2-47. This unit is dependent on the—

- Corps for legal, administrative, financial management, human resources, and transportation services, area damage control, CBRN decontamination assistance, mortuary affairs, and laundry and bath services.
- Quartermaster supply company or equivalent for Class I rations.
- Engineer company or equivalent for site selection, waste disposal, and minor construction.
- Movement control battalion or equivalent for supplemental transportation requirements.

ORGANIZATION AND FUNCTIONS

2-48. Figure 2-2 depicts the organization of the MMB. Specific information concerning the early entry and campaign elements is provided below.

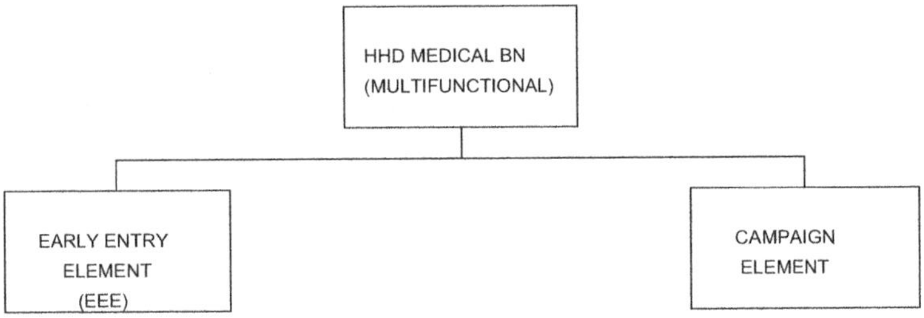

Figure 2-2. Multifunctional medical battalion

EARLY ENTRY ELEMENT, MULTIFUNCTIONAL MEDICAL BATTALION

2-49. Deployed units will begin operations immediately upon arrival in the theater avoiding the delays experienced in past conflicts. There is a need to deploy an EEE medical force to assess, plan, and coordinate AHS operations in the AO and provide medical surveillance, PVNTMED, and area medical support early on. One of the keys to providing early medical care is the development of modular medical elements. These modular AHS elements will perform specific medical functions and will allow medical planners to bring necessary medical units into theater as required by the battlefield mission.

MISSION

2-50. The mission of the EEE of the MMB, TOE 08486GA00, is to provide a scalable, flexible and modular C2, administrative assistance, logistical support, and technical supervision capability for assigned and attached medical organizations (companies, detachments, and teams) task-organized for support of deployed forces.

ASSIGNMENT

2-51. This EEE is assigned to an MMB.

EMPLOYMENT

2-52. Normally employed in a MEDCOM/MEDBDE AO. It can also be deployed to provide C2 to expeditionary forces in early entry operations and facilitates the reception, staging, onward movement, and integration (RSOI) of medical forces.

BASIS OF ALLOCATION

2-53. The BOA is one per MMB.

CAPABILITIES

2-54. At Level I, this unit provides—

- Command and control, staff planning, supervision of operations, medical and general logistics support as required, and administration of the assigned and attached units conducting AHS operations in its supported AO.
- Task organization of medical assets.
- Advice to senior commanders in the AO on the medical aspects of their operations.
- Coordination of medical regulating and patient movement within the AO.
- Monitoring, planning, and coordinating ground and air evacuation within the battalion AO. Coordinating air evacuation support requirements with the supporting aviation unit, and synchronizing the air evacuation plan into the overall medical evacuation plan.
- Consultation and technical advice on PVNTMED (medical entomology, medical and OEH surveillance, and sanitary engineering), COSC and BH, medical records administration, veterinary services, nursing practices and procedures, dental services, and automated medical information systems to supported units.
- Guidance for facility site selection and area preparation, as required.
- Monitoring and supervising medical logistics operations, to include Class VIII supply/resupply.
- Consolidated property book maintenance.
- Unit-level maintenance for wheeled vehicles and power generation equipment and wheeled vehicle recovery operations support to assigned or attached units.
- Organizational communications equipment maintenance support for the battalion.
- Food service support for staff and assigned/attached medical units.

DEPENDENCY

2-55. This unit is dependent on the—

- Corps for legal, administrative, financial management, human resources, religious support, transportation services, area damage control, CBRN decontamination assistance, mortuary affairs, and laundry and bath services.
- Quartermaster supply company or equivalent for Class I rations.
- Engineer company or equivalent for site selection, waste disposal, and minor construction.
- Movement control battalion or equivalent for supplemental transportation requirements.

ORGANIZATION AND FUNCTIONS

2-56. The battalion command section provides C2 of assigned and attached medical companies, detachments, and teams. (See Figure 2-3.)

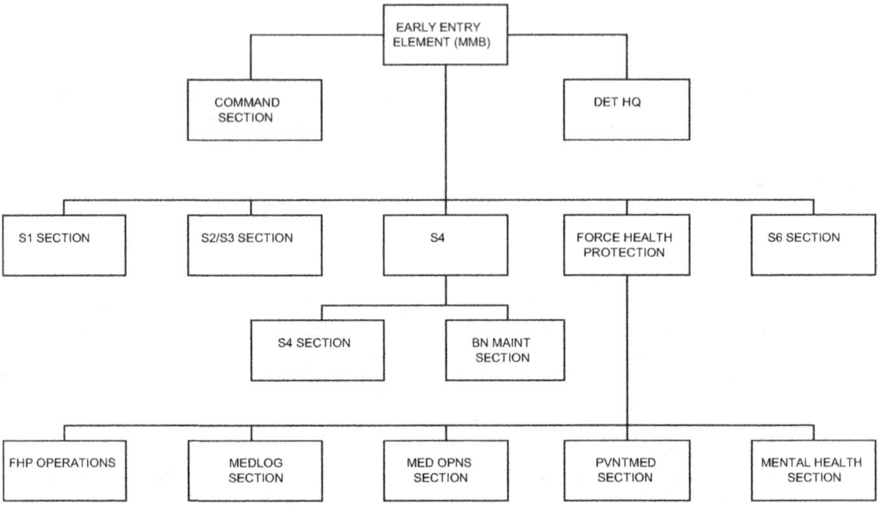

Figure 2-3. Early entry element, multifunctional medical battalion

2-57. The Adjutant (US Army) (S1) operates according to METT-TC and is the principle coordinating staff officer responsible for the delivery of human resources support. The S1 is responsible for the execution of all of the human resources (HR) core competencies to include: personnel readiness management (PRM); personnel accountability and strength reporting (PASR); personnel information management (PIM); reception, replacement, RTD, rest and recuperation (R&R), redeployment (R5) operations; casualty operations; essential personnel services (EPS); and postal operations. Morale, welfare, and recreation (MWR) operations, HR planning; and staff operations. The S1 also coordinates with elements of supporting agencies for finance, legal, religious, and administrative services.

2-58. The Intelligence Officer (US Army)/Operations and Training Officer (US Army) (S2/S3) section is responsible for security, plans and operations, deployment, relocation, and redeployment of the battalion and its assigned and attached units. The battalion's primary net control station is in this paragraph.

2-59. The Supply Officer (US Army) (S4) section is responsible for coordination, control, and management of logistics for assigned and attached units. The battalion maintenance section is under the staff supervision of the battalion S4. This section provides unit level maintenance for wheeled vehicles assigned to the HHD and assigned or attached units without unit level maintenance capability.

2-60. The FHP operations section is responsible for the planning, coordination, and execution of the AHS mission within the battalion's AO. This section supervises the operations of the medical logistics, operations, PVNTMED, and BH sections as well as coordinates medical regulation and patient movement within the AO. The medical operations section is responsible for—

- Management of area medical support (Roles 1 and 2).
- Planning, coordination, and execution of the Class VIII mission within the battalion's AO. This includes blood and medical maintenance management.

- Situational awareness for locations of ground and air ambulance assets.
- Planning and coordination for evacuation of patients using ground and air ambulance assets.
- Planning and coordination between air evacuation elements and the maneuver brigade S3. Information provided to the maneuver brigade S3 should include, but not be limited to the location of—
  - Medical treatment facilities, CCPs, and AXPs.
  - Forward air refueling equipment.
  - Location of supported units and liaison requirements. Air evacuation corridors, ground evacuation routes, and recommendations on usage.
- The PVNTMED section is responsible for the planning, coordination, and execution of the PVNTMED mission within the battalion's AO. This includes management of PVNTMED and veterinary assets.
- The BH section is responsible for the planning, coordination, and execution of the COSC mission within the battalion's AO. This section also collects and records social and psychological data.

2-61. The S6 section is responsible for all aspects of information management, automation, and communications-electronic (CE) support to assigned and attached units. The battalion's alternate net control station is in this paragraph.

2-62. The detachment headquarters provides for billeting, field feeding, discipline, security, training, and administration for personnel assigned to the EEE.

2-63. The battalion maintenance section is under the staff supervision of the battalion S4. This section provides unit level maintenance for wheeled vehicles assigned to the HHD and assigned or attached units without unit level maintenance capability.

CAMPAIGN SUPPORT ELEMENT, MULTIFUNCTIONAL MEDICAL BATTALION

MISSION

2-64. The mission of the campaign support element, MMB, TOE 08486GB00, is to complete the staffing of the headquarters of the MMB to enhance the delivery of campaign quality health care to deployed forces. To provide a scalable, flexible, and modular C2, administrative assistance, logistical support, and technical supervision capability for assigned and attached medical organizations (companies, detachments, and teams) task-organized for support of deployed forces.

ASSIGNMENT

2-65. This unit is assigned to an MMB.

EMPLOYMENT

2-66. Normally employed with the MEDBDE or a MEDCOM. If the EEE is deployed first, the campaign element falls in on this deployed module to provide a more robust C2 element and additional operational and planning capabilities, increased medical logistical staff, and a more diverse and robust clinical staff.

BASIS OF ALLOCATION

2-67. The BOA is one per MMB.

CAPABILITIES

2-68. At Role 1, this unit provides—

- Augmentation of C2 in personnel, logistics, HSS, FHP, medical operations, PVNTMED, BH, automation, maintenance, religious support, and food service to the EEE.
- Unit maintenance on all organic equipment except COMSEC equipment.

DEPENDENCY

2-69. This unit is dependent on the HHD, MMB for legal, administrative, financial management, human resources, food service, transportation services, area damage control, CBRN decontamination assistance, mortuary affairs, and laundry and bath services.

ORGANIZATION AND FUNCTIONS

2-70. The command section provides C2 for assigned and attached medical companies, detachments, and teams. (See Figure 2-4.)

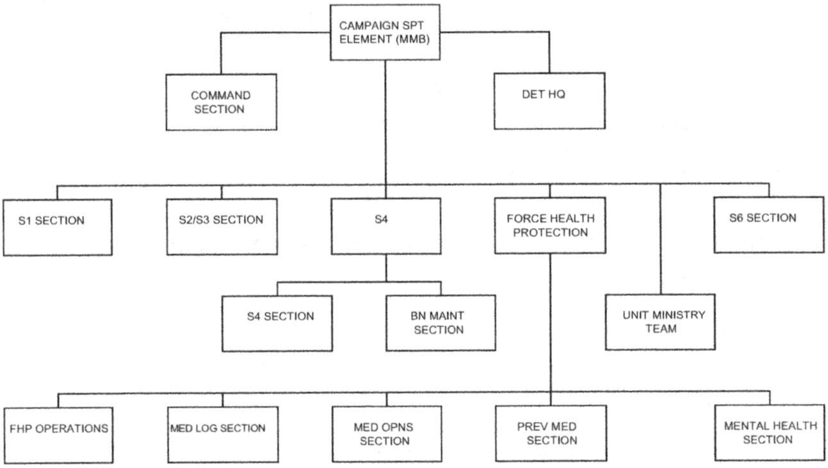

Figure 2-4. Campaign support element, multifunctional medical battalion

2-71. The S1 operates according to METT-TC and is the principle coordinating staff officer responsible for the delivery of HR support. The S1 is responsible for the execution of all of the HR core competencies to include: PRM, PASR, PIM, R5 operations, and casualty operations, EPS, postal operations, MWR operations, and HR planning and staff operations. The S1 also coordinates with elements of supporting agencies for finance, legal, religious, and administrative services.

2-72. The S2/S3 section coordinates with the EEE S2/S3 section of the MMB in the plans and operations support to subordinate organizational elements.

2-73. The S4 section is responsible for coordination, control, and management of logistics for assigned and attached units.

2-74. The FHP operations section plans and coordinates the HSS/FHP mission within the battalion's AO when collocated with the EEE. This section also provides advice on the operation of medical logistics, operations, PVNTMED, BH, medical regulating, and patient movement to supported organizations.

2-75. The medical logistics section plans and organizes Class VIII supply support (including blood and medical equipment management) in coordination with the medical logistics section of the EEE.

2-76. The medical operations section is responsible for—

- Management of area medical support (Roles 1 and 2).
- Situational awareness for locations of ground and air ambulance assets.
- Planning and coordination for evacuation of patients using ground and air ambulance assets.
- Planning and coordination between air evacuation elements and the maneuver brigade S3. Information provided to the maneuver brigade S3 should include, but not be limited to the location of—
  - Medical treatment facilities, CCPs, and AXPs.
  - Forward air refueling equipment.
  - Supported units and liaison requirements.
  - Evacuation corridors and evacuation routes and recommendations on usage.

2-77. The PVNTMED section plans and coordinates PVNTMED/veterinary support to subordinate organizations.

2-78. The BH section plans and coordinates BH support to subordinate organizations. This section also collects and records social and psychological data when collocated with the EEE.

2-79. The S6 section provides advice and support to subordinate organizations for CE support and information management.

2-80. The unit ministry team provides and coordinates religious support for all MMB staff, assigned/attached medical units and casualties in MMB subordinate unit MTFs.

2-81. The detachment headquarters provides coordination with the EEE in support of the field feeding and administration mission of the unit.

2-82. The battalion maintenance section provides unit level maintenance management support to assigned and attached units in coordination with the battalion S4 and the battalion maintenance section of the EEE.

SECTION V — GENERAL SUPPORT AVIATION BATTALION, COMBAT AVIATION BRIGADE

GENERAL SUPPORT AVIATION BATTALION

MISSION

2-83. The mission of the general support aviation battalion (GSAB) (Figure 2-5) is to provide the combat aviation brigade with aerial sustainment and maneuver support.

Figure 2-5. General support aviation battalion

ASSIGNMENT

2-84. Organic to the division combat aviation brigade, theater assault aviation brigade, theater general support aviation brigade, and the Army National Guard (ARNG) aviation expeditionary brigade (TOE 01300G400).

BASIS OF ALLOCATION

2-85. The BOA is one per division combat aviation brigade, one per ARNG aviation expeditionary brigade, one per theater assault aviation brigade, and three per theater GSAB (TOE 01300G400).

CAPABILITIES

2-86. This organization provides the following when all of its subordinate units are organized at Level I as shown in their respective TOEs:

- Battle command on the move.
  - Employment of Army airspace command and control (A2C2) system platforms in support of the division command group, maneuver, and aviation commanders' C2 requirements.
  - Radio relay and transport of liaison officers as required.
- Air assault.
  - Airlift for air assault operations and air movement of tactical forces (other than assault), field artillery or other fire support assets, tactical air defense (AD) systems, engineer equipment and personnel, and other resources.
  - Casualty evacuation when medical aircraft are not readily available.
  - Evacuation of downed aircraft and personnel.
- Sustainment.
  - Division sustainment with heavy-lift helicopter company.
  - General aviation support to the aviation brigade.
  - Support aircraft to the aviation support battalion.
  - Personnel recovery (PR).
  - Downed aircraft recovery team operations.
  - Casualty evacuation operations.
  - Air movement of general supplies, equipment, and personnel, including logistics over-the-shore (LOTS) when applicable.
  - Aerial noncombatant evacuation operations when directed by applicable commander/ authority.
  - Air movement of nuclear and other special munitions.
  - Self-deployment of organic CH-47 helicopters to a theater of operations when the aircraft have been configured with extended range fuel systems.

  - Retail level aircraft refueling when aircraft are configured with extended range fuel system (commonly referred to as "Fat Cow").
- Air traffic services support.
- Aeromedical evacuation.
  - Patient evacuation.
  - Patient movement between MTFs (patient transfers).
  - Class VIII resupply.
  - Joint blood program support.
  - Medical C2.
  - Movement of medical personnel and medical equipment.
  - Air crash rescue support.

2-87. Due to the CH-47's capabilities, heavy helicopter units can perform two unique missions:

- High-altitude operations.
- Oversized, heavy, and special munitions movement.

2-88. This GSAB is dependent upon the following organizations:

- Appropriate elements of the division aviation brigade for legal, HSS, FHP, financial management, and personnel and administrative services.
- United States Air Force weather team in the headquarters and headquarters company of the combat aviation brigade (TOE 01302G200) for air weather service support.
- Aviation support company (TOEs 01918G200 or 01919G000 [ARNG]) for intermediate level field maintenance support, DS maintenance for assigned aviation ground power units (AGPU), line item number (LIN) P44627, and support for aviation and common Class IX repair parts.

MEDICAL COMPANY, AIR AMBULANCE (HH-60), GENERAL SUPPORT AVIATION BATTALION, COMBAT AVIATION BRIGADE

OPERATIONAL INFORMATION

2-89. The medical company, air ambulance (Figure 2-6), provides aeromedical evacuation for all categories of patients consistent with evacuation precedence and other operational considerations within the division. In today's ongoing efforts in building a modular force, medical company, air ambulance will fall under the GSAB which will provide aircraft maintenance and logistics support, aviation communications, and real-time operational picture associated with today's combat environment.

MISSION

2-90. The mission of the medical company, air ambulance (HH-60) (TOE 08443G00) is to provide air ambulance evacuation support within the division and theater.

ASSIGNMENT

2-91. The medical company, air ambulance, (HH-60) is organic to the GSAB (TOEs 01125G000, 01225G100, and 01305G200) for C2.

EMPLOYMENT

2-92. This unit is employed in the division and corps area of responsibility. It is tactically located where it can best control its assets and execute its patient evacuation mission.

BASIS OF ALLOCATION

2-93. The BOA is one per GSAB and two per theater of operations (TOEs 01125G100, 01225G100, and 01305G100).

CAPABILITIES

2-94. At Level I, this unit provides—

- Twelve helicopter ambulances to evacuate critically wounded or other patients consistent with evacuation priorities and operational considerations, from points as far forward as possible, to division, corps, and EAC hospitals. Single patient lift capability for the company is 72 litter patients or 84 ambulatory patients, or some combination thereof.
- Four forward support medical evacuation teams (three helicopters each) that can be deployed as a team or an independent aircraft.
- Air crash rescue support.
- Expeditious delivery of whole blood, biological, and medical supplies to meet critical requirements.
- Rapid movement of medical personnel and accompanying equipment/supplies to meet the requirements for mass casualty (MASCAL), reinforcement/reconstitution, or emergency situations.
- Movement of patients between hospitals, ASFs, hospital ships, casualty receiving and treatment ships, seaports and railheads in the theater of operations.
- Military working dog evacuation and aeromedical evacuation support to a combat search and rescue task force.

DEPENDENCY

2-95. This unit is dependent upon—

- Appropriate elements of the division or corps for religious, legal, financial management, personnel and administrative services, and CE and COMSEC maintenance.
- The headquarters and headquarters company of the GSAB for religious, legal, HSS, FHP, personnel and administrative service, and unit CBRN support.
- The forward support company of the GSAB for Class III, automotive and generator maintenance, and field feeding.
- The aviation support company of the GSAB for unit level field maintenance of organic aircraft, including unit level supply support for aircraft Class IX.
- United States Air Force weather team in the headquarters and headquarters company of the division aviation brigade for air weather service support.

ORGANIZATION AND FUNCTIONS

2-96. The company headquarters provides C2, logistical support, and administration for the company.

2-97. The four forward support MEDEVAC teams (FSMTs) provide a task-organized means for medical evacuation at the DS and GS level. They also, provide emergency movement of medical personnel and emergency delivery of whole blood, biological, and medical supplies and equipment.

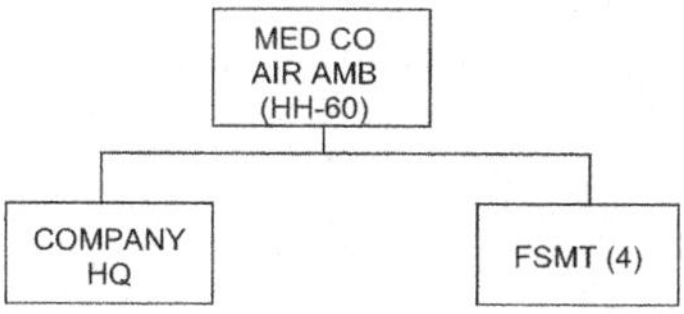

Figure 2-6. Medical company, air ambulance (HH-60)

AVIATION MAINTENANCE COMPANY, AVIATION UNIT MAINTENANCE, GENERAL SUPPORT AVIATION BATTALION, COMBAT AVIATION BRIGADE

MISSION

2-98. The mission of the aviation maintenance company is to perform aviation unit maintenance for the GSAB.

ASSIGNMENT

2-99. This unit is organic to a GSAB (TOE 01305G100).

EMPLOYMENT

2-100. The aviation maintenance company is employed as part of combined forces and provides unit level field maintenance support to the GSAB. The aviation maintenance company is located and its operations are normally conducted immediately behind the main battle area (MBA) in the EAB. The aviation maintenance company operations in forward areas are usually limited to daily inspections, minor adjustments, emergency and contact maintenance, and aircraft recovery. Contact teams and slice elements could be utilized as necessary.

BASIS OF ALLOCATION

2-101. The BOA is one per GSAB (TOE 01305G100).

CAPABILITIES

2-102. The aviation maintenance company provides unit level field maintenance support for 8 UH-60, 12 HH-60, and 12 CH-47 aircraft. Individuals of this organization can assist in the coordinated defense of the unit's area or installation. This unit performs unit level field maintenance on only the supported aircraft and some associated maintenance equipment.

DEPENDENCY

2-103. This unit depends on—

- Appropriate elements of the division or corps for religious, legal, HSS, FHP, financial management, personnel and administrative services, and logistical support (including supplemental transportation).
- Headquarters and headquarters company of the GSAB (TOE 01306G100) for religious, legal, HSS, FHP, financial management, personnel, and administrative services, unit-level logistical support, and communications maintenance.

- Forward support company (TOE 63317G200) for Class III, automotive and generator maintenance, and field feeding service support.
- Aviation maintenance battalion (TOE 01918G200), for unit level field maintenance support, backup unit level field maintenance support, DS maintenance for assigned AGPU, and support for aviation and common Class IX repair parts.

ORGANIZATION AND FUNCTIONS

2-104. This unit inspects aircraft systems, subsystems, components, and the maintenance performed by crew chiefs and repairers to determine airworthiness and compliance with applicable technical publications; performs operational checks and troubleshoots aircraft systems to isolate malfunctions; participates in maintenance test flights and maintains historical maintenance files.

2-105. This unit provides company maintenance operations. It maintains and regulates uniform flow of aircraft and associated components. Further, it maintains the company's aircraft logbooks, provides unit level Class IX support, and performs maintenance on aviation life support equipment (ALSE).

2-106. This unit provides unit level maintenance on UH/HH-60/CH-47 helicopters and associated equipment; services and lubricates helicopter systems and subsystems; and diagnoses and troubleshoots operational malfunctions.

MEDICAL COMMAND

2-107. The MEDCOM serves as the senior medical command within the theater, and works directly for the theater commander. With augmentation, MEDCOM is joint capable. In the absence of a theater commander the MEDCOM works for the regional combatant commander during a major combat operation (MCO). It possesses a centralized capability to effectively and efficiently task-organize medical elements based on theater specific medical requirements. The MEDCOM is depicted in Figure 2-7.

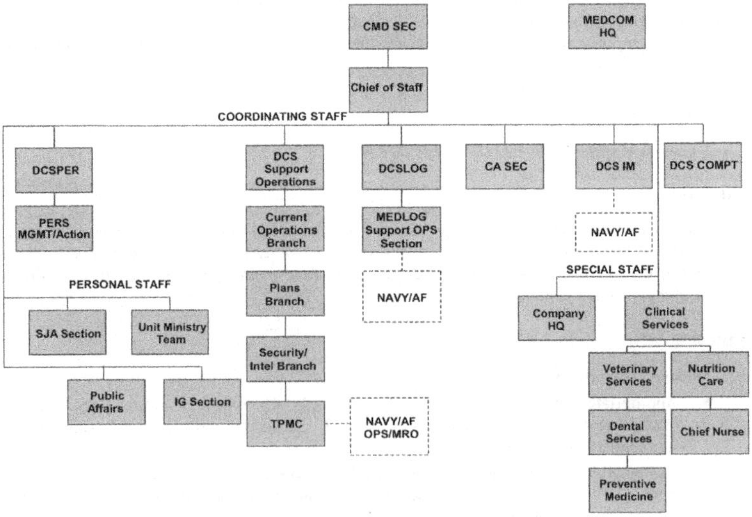

Figure 2-7. Medical command

2-108. The MEDCOM serves as the medical force provider (Figure 2-8) for the corps and EAC and provides the requisite medical C2 necessary to provide campaign-quality health care to deployed forces. The MEDCOM also focuses on theater medical operations plans and medical contingency plans.

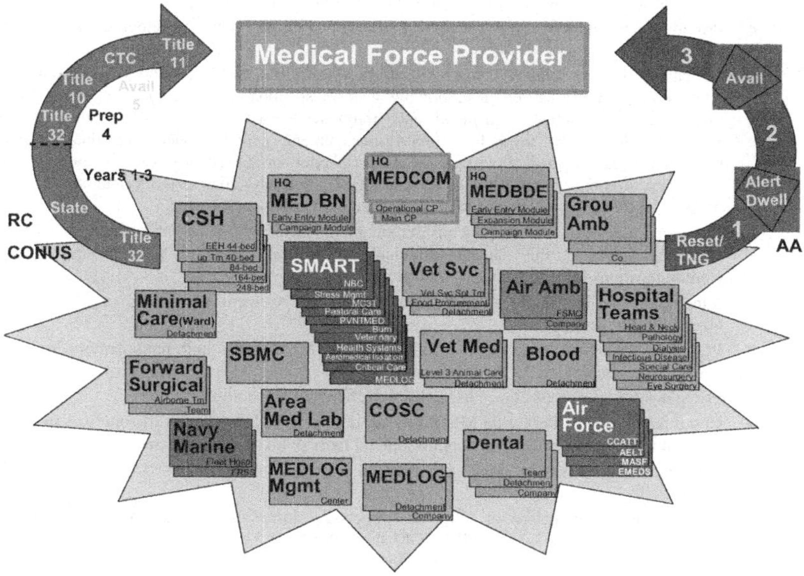

Figure 2-8. Medical force pool

2-109. The ability of the MEDCOM staff to develop accurate patient estimates based on real-time medical data ensures the right mix of medical specialties and resources are in the right place and at the right time to treat casualties on the battlefield. Care must be continuous and seamless from the POI or wounding through a phased system providing essential care in theater, with evacuation for definitive and rehabilitative care in CONUS.

2-110. The MEDCOM is a dedicated, regionally focused command with a basis of allocation of one per theater.

2-111. The MEDCOM is a versatile, modular medical C2 structure composed of a main command post (MCP) and an operational command post (OCP) which are 100 percent mobile. The MEDCOM is comprised of two standard requirements code (SRC) identified modules to provide a scalable medical C2 capability to the corps commander. Further, the MEDCOM is capable of—

- Providing subordinate medical organizations to operate under the subordinate MEDBDE and/or MMB and providing FSTs and/or other augmentation required by BCTs, and echelons above brigade (EAB) medical companies.
- Operating as the medical force provider for provisioning of required assets to the MEDBDEs.
- Advising the corps and other senior level commanders on the medical aspects of their operations.
- Providing staff planning, supervision of operations, and administration of assigned and attached medical units.
- Providing reach to the MHS.

- Providing medical Army support to other Services (ASOS)/Title 10 responsibilities of the ASCC commander.
- Coordinating with the USAF theater patient movement requirements center (TPMRC) for medical regulating and movement of patients from MEDCOM MTFs.
- Providing consultation services, technical advice, and clinical policy development in all aspects of medical and surgical services.
- Containing the functional staff to coordinate hospitalization, nursing, PVNTMED (entomology, epidemiology, OEH surveillance, potable water inspection, pest management, food facility inspection, and control of medical and nonmedical waste), medical evacuation, medical regulating, MEDLOG to include blood management, dental services, veterinary services (zoonotic disease control, investigation and inspection of subsistence, and animal medicine), pharmacy, optometry, nutrition care, and COSC, neuropsychiatric (NP), and BH services, medical laboratory services, and area medical support to supported units.
- Providing technical advice and consultation on medical automated information systems and programs such as the Theater Medical Information Program (TMIP) and Medical Communications for Combat Casualty Care (MC4).
- Functioning as the senior theater joint medical C2 element when augmented with joint augmentation package.
- Peacetime planning, training for and with the MEDBDEs, and participation in theater engagement plans.
- Providing advice and assistance in facility selection and preparation.
- Coordinating and orchestrating MEDLOG operations to include Class VIII supply, distribution, medical maintenance and repair support, optical fabrication, and blood management.
- Planning for the single integrated medical logistics manager (SIMLM), mission when designated.
- Planning and coordination of religious support for MEDCOM staff, assigned and attached medical units, and casualties in MEDCOM subordinate unit MTFs.

Operational Command Post

2-112. The MEDCOM OCP is a deployable, versatile module, which provides medical C2, policy development, and technical guidance to subordinate medical units and provides interface and liaison with supported theater forces in EAC, corps, and subordinate BCTs. The MEDCOM OCP has an assigned SRC to facilitate the placement of personnel in the CP and to integrate the CP into the time-phased force deployment list (TPFDL). The MEDCOM OCP is depicted in Figure 2-9.

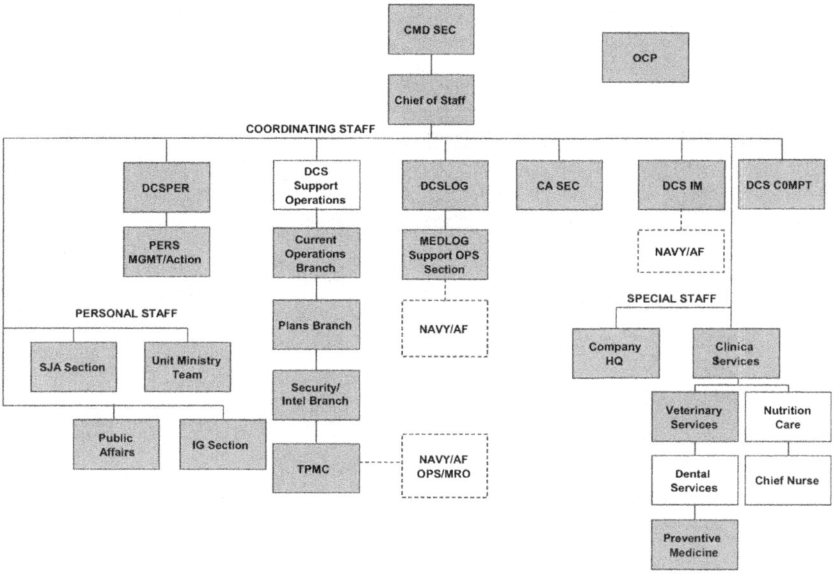

Figure 2-9. Operational command post, medical command

2-113. As the force continues to build within the theater, the MEDCOM's OCP can be incrementally expanded or augmented with additional planning and professional expertise. The deployment of personnel from the MCP model can then be built into the medical C2 required structure in order to provide campaign-quality health care to the deployed force.

Main Command Post

2-114. The MEDCOM MCP is a deployable augmentation module, which completes the staffing of the MEDCOM headquarters to enhance the delivery of campaign-quality health care to deployed forces in EAC, corps, and supported BCTs. It facilitates medical C2, policy development, and technical guidance to subordinate MEDBDEs and provides interface and liaison with supported theater forces in EAC, corps, and subordinate BCTs. If the OCP is deployed and needs additional personnel or clinical skills, the MEDCOM MCP can deploy personnel to the deployed OCP. This will provide a more robust medical C2 element and additional operational and planning capabilities; increased medical logistics staff; automated medical information systems; a more diverse clinical staff; and more depth in the command judge advocate (CJA), inspector general (IG), civil affairs (CA), and public affairs sections. The MEDCOM MCP has an assigned SRC to facilitate the placement on and the integration into the TPFDL. The MCP is depicted in Figure 2-10.

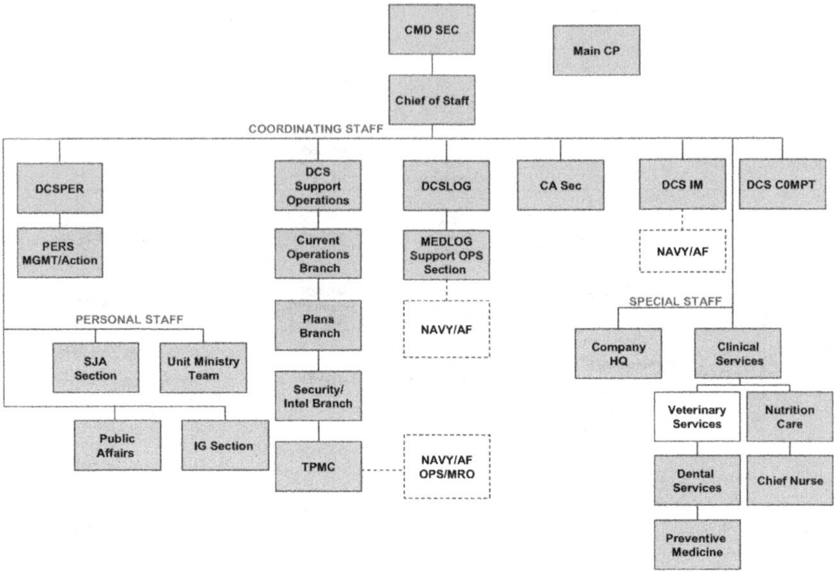

Figure 2-10. Main command post, medical command

MEDICAL BRIGADE

2-115. The MEDBDE provides a scalable expeditionary medical C2 capability for assigned and attached medical functional organizations task-organized in support of the deployed division/corps forces. The MEDBDE brings all requisite medical C2 and planning capabilities to provide responsive and effective HSS/FHP throughout the corps AO. The MEDBDE is depicted in Figure 2-11.

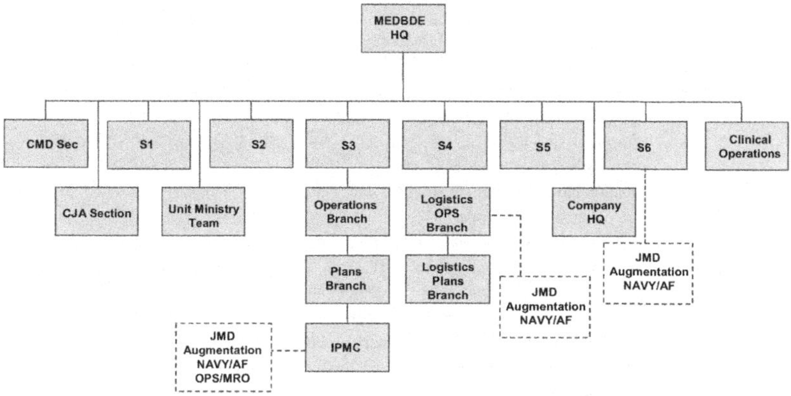

Figure 2-11. Medical brigade

2-116. The MEDBDE ensures the division/corps commander has the right mixture of medical professional (operational, technical, and clinical) expertise to synchronize the complex system of AHS functions required to maintain the health of his command by promoting fitness, preventing casualties from DNBI, and promptly treating and evacuating those injured on the battlefield.

2-117. The design and flexibility of the MEDBDE permits the division/corps commander the ability to meet expeditionary medical support requirements in support of early entry forces. As the supported corps commander's forces grow in both size and complexity, the MEDBDE can deploy additional modules that build upon one another to support full spectrum operations. The MEDBDE also provides the corps commander the appropriate medical C2 to continue to build his medical force capabilities through the integration of Army, joint, and multinational medical forces to ensure he can identify and counter any health threats in his supported areas. This will permit him to transition from expeditionary medical support operations to providing campaign-quality health care in his AO.

2-118. The MEDBDE has the capability to provide SRC elements for an early entry module (EEM), an expansion module, and the campaign module (CM), thus enabling its capability to be tailored to METT-TC factors of a specific division/corps operation.

2-119. The MEDBDE provides medical C2, administrative assistance, and technical supervision of assigned and attached medical units providing AHS in support of a division/corps. Further, it is capable of—

- Providing a rapidly responsive early entry medical C2 module that can quickly integrate into the early entry deployment sequence for crisis management.
- Providing full spectrum continuous medical C2 in support of all Army corps and joint (when augmented) forces.
- Providing operational medical augmentation to Role 2 BCT medical companies.
- Advising corps and BCT commanders on the medical aspects of their operations.
- Providing medical staff planning, operational and technical supervision, and administrative assistance for MMBs and CSHs task-organized under the MEDBDE.
- Coordinating with supporting TPMRC for medical regulating and medical evacuation from MMBs and hospitals to supporting EAC MTFs and CONUS.

- Providing medical consultation services and technical advice in—
  - Preventive medicine (medical and OEH surveillance, environmental health, sanitary engineering, and medical entomology).
  - Nursing services.
  - Dental services.
  - Behavioral health (to include COSC) and NP care.
  - Veterinary services (including food safety and inspection, animal medical care, and veterinary PVNTMED services).
  - Nutrition care and medical food service.
  - Medical laboratory support.
- Providing advice and recommendations for conduct of civil-military operations (CMO).
- Control and supervision of Class VIII supply and resupply movement to include blood management. When designated by the combatant command, serves as the SIMLM.
- Providing joint capable medical C2 capability when augmented with appropriate joint assets.
- Serves as the executive agent for veterinary services according to DODD 5200.31E.
- Coordinates ASOS for ship-to-shore/shore-to-ship medical evacuation mission.
- Plans and coordinates religious support for MEDBDE staff, assigned and attached medical units, and casualties in MEDBDE subordinate unit MTFs.

Chapter 3

Operational and Tactical Evacuation Planning

A comprehensive medical evacuation plan is essential to ensure effective, efficient, and responsive medical evacuation is provided to all wounded, injured, and ill Soldiers in the JOA. The Army medical evacuation plan flows from the GCC's guidance and intent and incorporates all missions and tasks directed by the GCC to be accomplished and is synchronized with supporting and supported units. In some scenarios, Army air and ground evacuation resources may be directed to provide support to sister Services, allied and coalition partners, and HN forces. Examples of a medical evacuation OPLAN and OPORD are contained in Appendix D.

SECTION I — THEATER MEDICAL EVACUATION PLANNING RESPONSIBILITIES

JOINT PLANNING

3-1. All medical evacuation planning within the theater reflects the GCC's guidance and intent. Planning at the joint level is governed by Joint Publications (JPs) 5-0, 5-00.1, and 5-00.2. Guidance on medical evacuation in joint operations is addressed in JP 4-02. When directed by the GCC, Army medical evacuation assets may be tasked to support other than Army forces engaged in the execution of the joint mission. These additional support missions will be clearly articulated in the GCC's OPLAN and OPORD. The ASCC surgeon, with the advice of the senior medical evacuation planner, will coordinate and synchronize these support operations with the combatant command surgeon, joint task force (JTF) surgeon, and the other Services and/or multinational partners as required to ensure that a comprehensive and effective, efficient, and responsive plan is developed and implemented.

MEDICAL COMMAND AND CONTROL ORGANIZATIONS

3-2. Army evacuation assets support the full spectrum of all Army operations and are capable of supporting Army forces engaged in the execution of the joint mission, as directed. Major combat operations supported by an ASCC will have a deployed MEDCOM. The C2 coordination and orders flow for AHS is illustrated in Figure 3-1. Evacuation planning should occur at each major level of command in order to effectively execute command, control, and influence over evacuation planning.

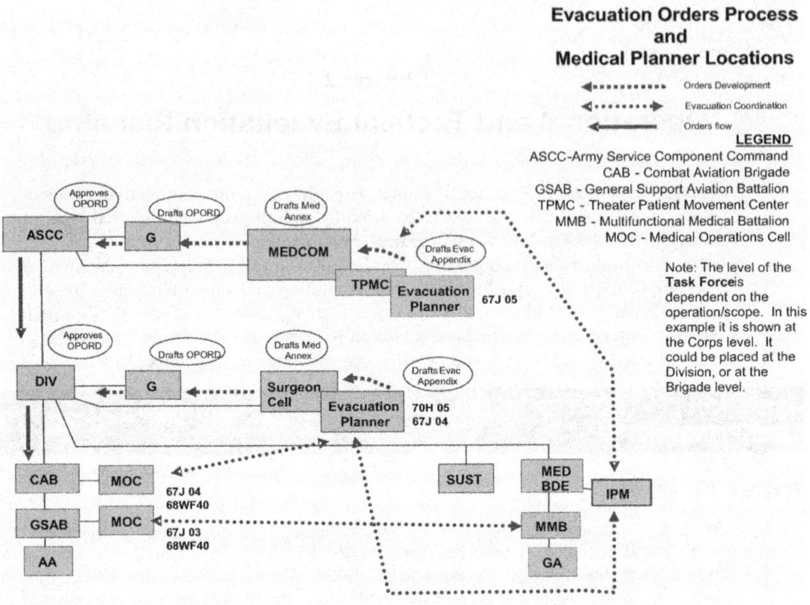

Figure 3-1. Command and control organizations for coordination and orders flow

3-3. Under the new Army modular design, some early entry operations may not require the deployment of a theater or a brigade C2 structure. In these cases, medical evacuation planning, coordination, and synchronization are initiated by the most senior medical command/staff level (such as the division surgeon, medical brigade, GSAB, and/or MMB). The senior leaders responsible for air and ground medical evacuation provide estimates and recommendations for the plan. This plan is normally provided as an appendix to the HSS annex and is disseminated down through the chain of command.

3-4. Dedicated Army rotary-wing aircraft (air ambulances) are under the C2 of the GSAB within the transformed Army aviation structure. Ground ambulances remain under the C2 of medical units/elements that are organic to movement and maneuver, fires, and sustainment units. The medical company, ground ambulance is under the C2 of the MMB.

3-5. To ensure an integrated medical evacuation system, synchronization of air with ground ambulance assets is essential. To affect this synchronization, the medical C2 headquarters must maintain influence over the execution of the medical evacuation plan. This influence is accomplished through the orders process.

Theater Patient Movement Center

3-6. In the design of the modular MEDCOM, the medical reengineering initiative (MRI) MEDCOM's medical regulating office (MRO) staff was reevaluated, realigned, and augmented with medical evacuators to more effectively coordinate and synchronize the various ground and air evacuation aspects of the operation and to supervise and facilitate the medical regulating functions. This staff section was renamed as the theater patient movement center (TPMC). The addition of medical evacuators enables this section to develop, coordinate, and synchronize the medical evacuation portion of the MEDCOM AHS

hour coordination and management responsibility for patient regulating and administration within the MEDCOM AO and directs patient evacuation between facilities within its AO as required to avoid patient overflow. This staff is responsible for preparing patient statistical reports for the command. This section also coordinates with the supporting TPMRC for the transfer of patients out of the MEDCOM/theater. The TPMRC regulates the flow of patients out of the theater to an MTF capable of providing the required specialty care and arranges for the transport of these patients by USAF aeromedical evacuation assets.

INTRATHEATER PATIENT MOVEMENT CENTER

3-7. As with the modular MEDCOM, the MRO of the modular MEDBDE was also redesigned in a similar fashion to enhance the capabilities for medical evacuation planning within the MEDBDE AO. This staff section was renamed the intratheater patient movement center (IPMC). This office is responsible for coordinating and synchronizing medical evacuation operations with the TPMC and the division surgeon cell. It regulates patients within its subordinate hospitals and arranges for transportation to transfer patients between facilities to relieve surgical backlogs, to obtain specialty care, and to ensure bed availability prior to start of operations, as appropriate. It also coordinates with the MEDCOM TPMC and the TPMRC for the evacuation of patients out of theater. This office maintains a 24-hour coordination and management responsibility for patient regulating and administration with the MEDBDE AO. It also prepares patient statistical reports for the MEDBDE commander. In the absence of the TPMC, IPMC assumes duties of the TPMC.

SECTION II — PLANNING PROCESS

EVACUATION PLANS AND ORDERS

CHARACTERISTICS

3-8. Operation plans and orders are the means by which commanders express their visualization, commander's intent, and decisions. They focus on results the commander expects to achieve. These OPLANs and OPORDs provide the basis for ensuring that medical evacuation operations are comprehensively planned, synchronized, responsive, and ensure a seamless continuum of care.

3-9. Evacuation plans and orders help form the basis commanders use to synchronize military operations. They encourage initiative by providing the *what* and *why* of a mission, and leave the *how to accomplish the mission* to subordinates. They give subordinates the operational and tactical freedom to accomplish the mission by providing the minimum restrictions and details necessary for synchronization and coordination. As medical evacuation operations are complex and may cross command and service boundaries, it is essential that the medical evacuation annex to the medical OPLAN/OPORD clearly delineates responsibilities, communications interoperability and procedures, and coordination requirements.

PLANS AND ORDERS

3-10. Publishing the commander's OPLAN/OPORD permits subordinate commanders to prepare supporting plans and orders. They can discern and implement instructions derived from a higher commander's plan and order. Appendix D provides an example of operational planning and a sample OPORD. Additionally, the commander's OPLAN/OPORD—

- Focuses subordinates' activities.
- Provides tasks and activities, constraints, and coordinating instructions necessary for mission accomplishment.
- Encourages agility, speed, and initiative during execution.
- Conveys instructions in a standard, recognizable, clear, and simple format.

3-11. The amount of detail provided in an evacuation plan or order depends on several factors, to include the experience and competence of subordinate commanders, cohesion and tactical experience of subordinate units, and the complexity of the operation.

3-12. Commanders balance these factors with their guidance and commander's intent, and determine the type of plan or order to issue. To maintain clarity and simplicity, plans and orders include annexes only when necessary and only when they pertain to the entire command. Annexes contain the details of support and synchronization necessary to accomplish the mission.

3-13. For a complete discussion of the military decision-making process and development of OPLANs/OPORDs refer to FM 5-0 and FMI 5-0.1.

Medical Evacuation Planning

3-14. Medical evacuation planning which supports the HSS/FHP plan is an ongoing process and is accomplished by medical operations element at battalion, brigade, division, corps, and theater levels. The HSS/FHP operations planning process must take under consideration all issues that could affect or influence HSS/FHP support for tactical operations. These issues should be considered in the initial developmental stages of theater, corps, division, or brigade plan as the supporting HSS/FHP annex is developed.

3-15. Planning is a dynamic and continuous process. The framework established in a robust initial plan can be built on in subsequent phases of an operation. The evacuation plan for a given AO will require adjustment and change during different phases of an operation. For example, transitioning from offensive operations to stability operations will generally dictate a significant change in evacuation coverage priorities.

3-16. Medical evacuation planning considerations, although essentially the same at all levels of command, may entail a broader scope, complexity, and detail at the higher levels of command due to the coordination and synchronization requirements which exist across command lines and service or multinational boundaries. Table 3-1 provides medical evacuation planning considerations at various levels of command based on METT-TC.

Table 3-1. Medical evacuation planning considerations

Mission	1. Define the protocol for a valid evacuation mission (mission statement). This protocol will serve as the authority to execute the mission by each medical evacuation unit/element/crew. 2. Use the HSS/FHP annex and mission analysis to: • Task evacuation units where needed. This may involve tasking part or all of DS units to perform a GS or area support mission. 3. Based on the type of operation: • Define trigger points for changes in the evacuation support plan. • Develop a concept of operations for all of the corps evacuation resources. 4. Define task organization of units assigned to the corps (and when directed to the JTF) and support relationships, if possible. Lines of coordination and mutual support relationships should be clearly delineated. 5. Area support and GS coverage areas will be

	defined with flexibility built into the plan. Coverage areas along division boundaries, for example, should not restrict or hinder evacuation of neighboring units that are geographically close. 6. Define pilot certifications for shore-to-ship medical air evacuation. 7. Develop standing operating procedure (SOP) requirements for single versus dual ship medical air evacuation. Address theater-specific procedures for in-flight link-up between medical evacuation aircraft and nonmedical evacuation aircraft for the purpose of dual ship requirements and/or security of pick-up zone. 8. Delineate any nonstandard air evacuation missions and allocate resources accordingly, such as support to Army special operations, joint and coalition forces. 9. Coordinate with any units that will have shared responsibilities for evacuation in a specific area and/or a specific mission. 10. Evaluate medical evacuation specific A2C2 considerations. Specifically, coordinating the crossing of brigade/division/corps boundaries/ area of operations. Medical evacuation must be afforded special expedited procedures for medical evacuation missions. Additionally, METT-TC dependent air corridors may have to be developed for the medical evacuation mission.
Enemy	1. Define the level or air-to-air threat and the threat faced by ground evacuation resources and the impact of this threat on the overall evacuation plan. This will be a major consideration in dictating which means of evacuation will be the primary means. 2. Based on the estimate of the number of enemy prisoners of war (EPW) or other categories of detainees involved in the operation— • Define procedures for evacuating wounded, injured, or ill persons through medical channels to include responsibilities of the echelon commander for designating guards for those evacuated directly from the point of capture to an MTF. • Allocate assets as necessary to provide this support and to support detainee collection points and holding areas.
Terrain and Weather	Define the types of terrain in the theater and use information as a major consideration in dictating the primary means of evacuation (ground or air in

	each sector) and any factors that may impair evacuation efforts.
Troops and Support Available	1. Disseminate locations of facilities used in evacuation, MTFs, MASFs, and aerial ports of debarkation (APOD). 2. Disseminate the capabilities and availability of joint evacuation assets and how to request their assistance. 3. Collect contact information from all evacuation units, assemble and publish for theater use. 4. Standardize a mandatory evacuation mission data collection format and collection method and schedule.
Time Available	1. Define the acceptable limits of evacuation time based on the distance from evacuation units. 2. Determine the proper allocation of resources to support the entire theater requirements. 3. Plan for trigger points (in time or distance) for changes in the evacuation plan to occur. 4. Define phases of the operation that reflect the commander's intent. Major changes in evacuation coverage usually occur when changes occur in the operational tempo (OPTEMPO). 5. Define briefing levels for launch authority. Consider developing a uniform theaterwide aviation risk assessment sheet.
Civil Considerations	1. Based on the HSS/FHP estimate and plan, determine requirements for the evacuation of HN civilians and others. 2. Define and disseminate protocols and procedures for evacuating civilians. 3. Allocate resources to support this mission, when directed.

3-17. For an effective HSS/FHP plan, medical staffs must coordinate and share operations information in order to synchronize support operations for mission accomplishment. See FMs 4-02, 4-02.1, 4-02.6, 4-02.12, 4-02.17, 4-02.25, 8-42 (4-02.42), and 8-55 (4-02.55), for doctrinal guidance on HSS/FHP operations. After the HSS/FHP plan is completed, it is incorporated into the C2 headquarters plan. After all annexes and appendixes of the plan are approved by the commander, it is incorporated into the OPORD.

SURGEON'S RESPONSIBILITIES

3-18. The surgeon is responsible for supervision and development of AHS input for the OPORD. The HSS plan serves as the base document for this input. The HSS plan is revised or updated based on mission analysis or changes in HSS requirements. The surgeon is tasked by the Assistant Chief of Staff (Operations and Plans) (G3)/S3 for HSS/FHP input to the OPORD for support of operations. The G3/S3 indicates timeline requirements. The surgeon and his staff are involved in all stages of the planning process, to identify all HSS requirements. Information on development of running estimates and the OPLAN/OPORD are discussed below. The medical plan/operations cell develops a HSS/FHP plan based on guidance received from the commander and the surgeon and provides HSS/FHP operational planning updates to the surgeon. The plan is briefed to the commander for approval, as required and provided to the G3/S3 according to the format of the in FM 8-55 (4-02.55).

MEDICAL EVACUATION TOOLS

3-19. It is essential that the evacuation plan for all combat operations be well planned, coordinated, and disseminated. In designing the medical evacuation plan, the medical planner should consider the following:

- Casualty collection points normally are predesignated along the axis of advance or evacuation routes.
- Forward of the BAS, the trauma specialist, CLS, and combat troops take casualties to the CCPs. These points facilitate acquisition by supporting ambulance teams and reduce evacuation time.
- When used by the BAS, CCPs help preserve BAS mobility, preclude carrying casualties forward, and reduce evacuation time to the sustainment area. When designating a CCP, the designating authority makes a decision whether or not to provide medical staff at the location. This decision is based upon the assessment of risk versus the availability of personnel.
- Normally, the role of care designating the point is responsible for staffing. Medical personnel may not be available to staff these points, and CLSs and ambulatory patients may be required to perform self-aid, buddy aid, or enhanced first aid. Casualty collecting points should be identified on operational overlays.

Note. A CCP staffed by a trauma treatment team is designated as a BAS (minus) rather than as a CCP.

3-20. Ambulance exchange points are a predetermined position where patients are exchanged from one evacuation platform to another.

3-21. These points are normally preplanned and are a part of the HSS/FHP annex to the OPLAN. In the forward area, the threat of enemy ground activities, large concentrations of lethal weapons systems, and effective use of antiaircraft weapons may dictate that the AXP be a predetermined rendezvous point for the rapid transfer of patients from one evacuation platform to another. The location of AXPs should be frequently changed to preclude attracting enemy fires.

3-22. Ambulance exchange points are established for many different reasons:

For example, the ambulance platoon of the heavy BSMC possesses a mixture of wheeled and tracked ambulances.

- The brigade tracked ambulances are provided so that they may keep up with maneuver elements. These vehicles carry the patients from the BAS to an AXP where the brigade wheeled ambulances take over for the relatively longer trip to the sustainment area.
- Ambulance exchange points are not limited to ground evacuation assets. Another example is a situation where the threat air defense artillery capability is such that air ambulances cannot fly as far forward as the BASs. However, an AXP could be established a few kilometers to the rear, still well forward of the BSA.
- The brigade combat team (BCT) tracked or wheeled ambulances could then transfer the patients to the air assets, thereby facilitating the rapid evacuation of patients and realizing a significant time savings.

3-23. By using AXPs evacuation assets are returned to their supporting positions faster. This facilitates evacuation as the returning crews are familiar with the road network and the supported unit's tactical situation. In the case of air ambulance (AA) assets, it is important because of the requirements for integration into the A2C2 system at each role and the enhancement to survivability provided by current threat and friendly air defense information.

3-24. The ambulance shuttle system is an effective and flexible method of employing ambulances during combat. It consists of one or more ambulance loading points, relay points, and when necessary, ambulance control points, all leveled forward from the principal group of ambulances, the company location, or basic relay points as tactically required. The ambulance—

- Loading point is a point in the shuttle system where one or more ambulances are stationed ready to receive patients for evacuation.
- Relay point is a point in the shuttle system where one or more empty ambulances are stationed. They are ready to advance to a loading point or to the next relay post to replace an ambulance that has moved from it. As a control measure, relay points are generally numbered from front to rear.
- Control points consists of a Soldier (from the ambulance company or platoon) stationed at a crossroad or road junction where ambulances may take one more directions to reach loading points. The Soldier, knowing from which location each loaded ambulance has come, directs empty ambulances returning from the sustainment area. The need for control points is dictated by the situation. Generally, they are more necessary in forward areas.

3-25. In the establishment of the ambulance shuttle system, once the relay points are designated, the required numbers of ambulances are stationed at each point. If the tactical situation permits, the ambulances may be delivered to the relay points by convoy.

3-26. Staffing of relay, loading, and ambulance control points. Important points may be manned to supervise the blanket, litter, and splint exchange and to ensure that messages and medical supplies to be forwarded are expedited.

3-27. Advantages of the ambulance shuttle system are that the system—

- Places ambulances at CCPs and BASs as needed.
- Permits a steady flow of patients through the system to MTFs.
- Avoids unnecessary massing of transport in forward areas.
- Minimizes the danger of damage to ambulances by the enemy.
- Permits the commander or platoon leader to control his elements and enables him to extend their activities without advancing the headquarters.
- Facilitates administration and maintenance.
- Maximizes the use of small battle command elements (sections or platoons) to operate the ambulance shuttle without employing the entire parent unit.
- Provides for flexible use of other ambulance assets for specific situations.

OBSTACLES MARKING

3-28. Ambulance crews must know and recognize the standard land/gap marking patterns. Unit TSOPs on the types of materials used should be available to ambulance crews. Refer to FM 3-34.2 for additional information.

Chapter 4

Army Medical Evacuation

Medical evacuation operations are planned to provide comprehensive, responsive, flexible, and agile support to the tactical commander in conformity with the commander's intent and OPLANs. This chapter discusses the employment of medical evacuation resources and the coordination and synchronization required to effectively execute medical evacuation operations (to include the transfer of patients between MTFs and to staging facilities) by air and ground evacuation assets.

SECTION I — MEDICAL EVACUATION SUPPORT

EVACUATION PRECEDENCE

4-1. Casualties requiring evacuation are prioritized to ensure the most seriously injured or ill receive timely medical intervention consistent with their medical condition. As with medical treatment, medical urgency is the only factor used to determine the medical evacuation precedence. (Appendix A provides an in-depth discussion of the provisions of the Geneva Conventions.)

This paragraph implements STANAG 3204.

4-2. The determination to request medical evacuation and assignment of a precedence is made by the senior military person present or, if available, the senior medical person at the scene. This decision is based on the advice of the senior medical person at the scene (if available), the patient's condition, and the tactical situation. Assignment of a medical evacuation precedence is necessary. The precedence provides the supporting medical unit and controlling headquarters with information that is used in determining priorities for committing their evacuation assets. For this reason, correct assignment of precedence cannot be overemphasized; over classification may result in an increase in evacuation which could burden the HSS system.

4-3. The patient's medical condition is the overriding factor in determining the evacuation platform and destination facility. The AA operates wherever needed on the battlefield, dependent on risk and METT-TC factors. The crew of the AA, assisted by onboard patient monitoring and diagnostic equipment, is trained in aeromedical procedures to provide optimum en route patient care. It is the preferred method of evacuation for most categories of patients. Air ambulances are a low-density, high-demand resource and must be managed accordingly. To conserve these valuable resources, medical planners should plan to use AA to primarily move Priority I, URGENT and Priority IA, URGENT-SURG patients with the other categories on a space-available basis. Health service support planners must plan for a synchronized air and ground evacuation plan. Depending on the length of time required for an AA to be dispatched and arrive at the POI, it may be prudent to evacuate the casualty by ground evacuation assets to a BAS and/or Role 2 MTF for stabilization by a physician. Further evacuation could then be accomplished by AA.

4-4. Patients will be picked up as soon as possible, consistent with available resources and pending missions. Table 4-1 depicts the categories of evacuation precedence and the criteria used to determine the appropriate precedence.

Table 4-1. Categories of evacuation precedence

Priority I—URGENT	Is assigned to emergency cases that should be evacuated as soon as possible and within a maximum of 2 hours in order to save life, limb, or eyesight, to prevent complications of serious illness, or to avoid permanent disability.
Priority IA—URGENT-SURG	Is assigned to patients who must receive far forward surgical intervention to save life and to stabilize them for further evacuation.
Priority II—PRIORITY	Is assigned to sick and wounded personnel requiring prompt medical care. This precedence is used when the individual should be evacuated within 4 hours or his medical condition could deteriorate to such a degree that he will become an URGENT precedence, or whose requirements for special treatment are not available locally, or who will suffer unnecessary pain or disability.
Priority III—ROUTINE	Is assigned to sick and wounded personnel requiring evacuation but whose condition is not expected to deteriorate significantly. The sick and wounded in this category should be evacuated within 24 hours.
Priority IV—CONVENIENCE	Is assigned to patients for whom evacuation by medical platform is a matter of medical convenience rather than necessity.
The NATO STANAG 3204 has deleted the category of Priority IV—CONVENIENCE; however, it will still be included in the US Army evacuation priorities as there is a requirement for it on the battlefield.	

SECTION II — MEDICAL EVACUATION REQUESTS

4-5. This section discusses requests for medical evacuation. Specific procedures, frequencies, and security requirements for transmittal of medical evacuation requests are delineated through the orders process and are made a part of the unit/command SOPs. Each sector based on the METT-TC may be designated with a different method of evacuation as the primary means to effect evacuation. In sectors which have a high ground-to-air or air-to-air threat may rely on ground evacuation assets to move the majority of patients. In other sectors where the ground threat is high and comprised of small arms, improvised explosive devices, and bombs, medical evacuation operations may be more efficiently and effectively executed by AAs. An additional consideration in planning medical evacuation operations is to determine whether armed escorts are required for either the ground or AA mission. Those missions that require armed escort must be thoroughly coordinated and synchronized between the medical assets and force protection assets that will accompany them.

4-6. Medical evacuation requests often are sent from the POI, through intermediaries, such as higher headquarters, who then transmit the request up to the nearest medical evacuation unit. The unit relaying the request must ensure that it relays the exact information originally received. The radio call sign and frequency relayed (Line 2 of the request) should be that of the requesting unit and not that of the relaying unit. However, the intermediaries contact information can be given as additional information if a callback is necessary to clarify details of the mission. Figure 4-1 depicts the communication flow for evacuation requests.

4-7. The necessity for secure communication in transmitting a 9-line request is METT-TC dependent. A technically capable enemy may be able to intercept a nonsecure 9-line request and strike at the pickup zone causing additional harm to the patients and the medical evacuation crew. In combat and when directed by higher headquarters, always use secure communications when requesting medical evacuation.

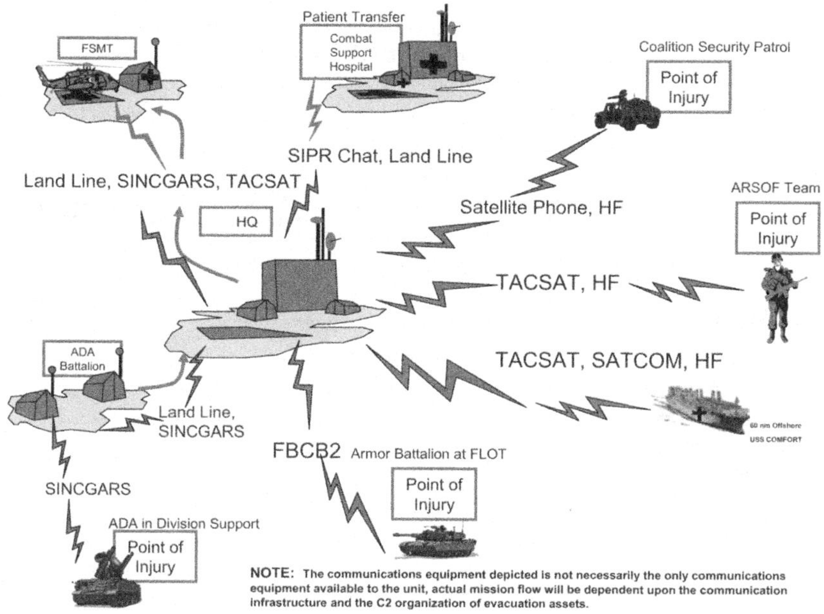

Figure 4-1. Flow of communication for evacuation request

GROUND EVACUATION REQUEST

4-8. Figure 4-2 provides an example of a request from the POI for ground evacuation. The patient is a ROUTINE precedence patient requiring evacuation from the POI to the supporting BAS (Role 1 MTF) with follow-on evacuation to an AXP for further evacuation to the rear. Should the patient require further treatment, either an air or ground evacuation could be used.

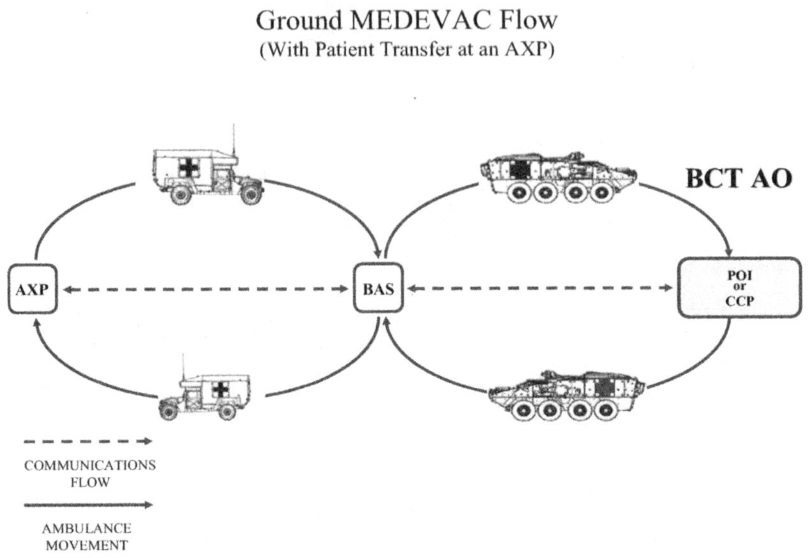

Figure 4-2. Ground evacuation request in a maneuver unit

4-9. Ground medical evacuation assets also provide DS and GS to supported units in the EAC. Figure 4-3 and Figure 4-4 depict the processes of providing DS to an ASMC and the GS support role of transferring a patient from a corps hospital to a MASF for further evacuation out of the theater.

4-10. In the corps area, the ASMC provides Roles 1 and 2 medical support on an area basis. The ASMC has organic evacuation assets and receives augmentation when required from the medical company ground ambulance assigned to the MMB.

4-11. Medical evacuation resources are also used to transfer a patient between hospitals within the theater and from a hospital to MASF for further evacuation out of theater. In Figure 4-3, this mission is being accomplished by a ground ambulance. However, depending upon the distance from the hospital to the MASF and/or APOD, this mission can also be accomplished by AA assets.

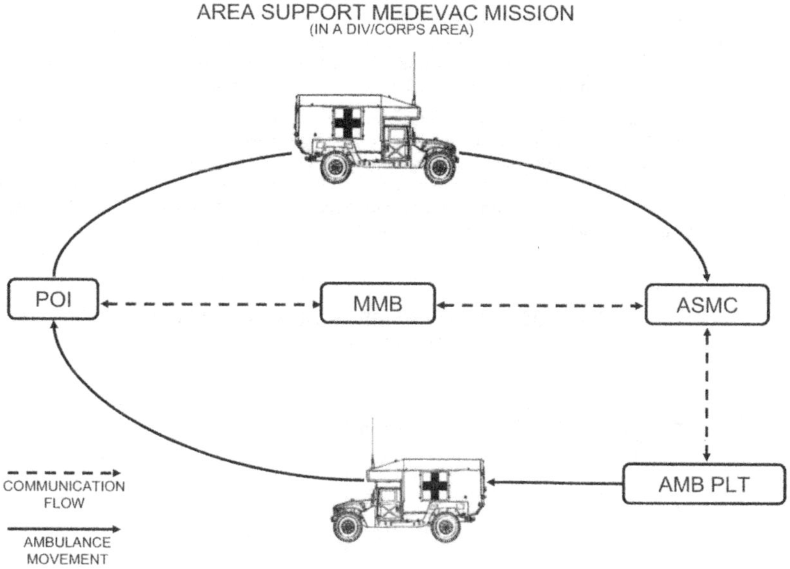

Figure 4-3. Ground evacuation request in the corps

AIR EVACUATION REQUEST

4-12. Air ambulances may fly as far forward as possible on the battlefield. Although evacuation by AA is the preferred means for all casualties, when high evacuation workloads exist, evacuation by AA should be the primary means used for URGENT and URGENT-SURG patients. Figure 4-5 depicts the zones of evacuation. The GSAB/CAB will place AA assets where they can best support the tactical commander's plan and facilitate the timely and responsive evacuation. To coordinate and synchronize the ground and air medical evacuation plans, a staff element was designed within the GSAB headquarters. This cell consists of the medical planners on the GSAB and CAB staff. The medical operations cell (MOC) will:

- Establish flight procedures specific to air evacuation missions within the CAB. This may include special routes or corridors as well as procedures for escort aircraft linkup.
- Ensure lines of communication (LOC) to supported units and higher levels of medical command are available. Further the MOC ensures supported units understand medical evacuation procedures and capability.
- Establish medical evacuation briefing and launch procedures. Ensure there is 24-hour access to those able to launch high and very high-risk missions.
- Maintain awareness of tactical and medical situation. Coordinate with MROs at higher levels to efficiently conduct the general support mission and to work in concert with adjacent units.
- Assist the AA CO and GSAB/CAB staff conduct medical evacuation operations.

- Coordinate missions with supported command surgeons. To ensure situational awareness and coordination of medical evacuation efforts, this staff keeps command surgeons informed on air evacuation missions being executed in their AO.
- Consults and coordinates with supported command surgeons when AAs cannot be launched to execute a requested mission. Ensures that the appropriate command surgeon is notified to ensure mission can be accomplished by ground evacuation assets.

4-13. Figure 4-4 depicts a request for air medical evacuation. The scenario shows the request being sent to the FSMT, the AA company, and the GSAB. The routing of the request would be dependant on how our forces are arrayed. In a similar situation where ground evacuation may not be readily available or the tactical situation precludes its use the requesting unit could transmit their medical evacuation request directly to a supporting FSMT as a primary means for a medical evacuation request or the AA company as a secondary option with a third being sent directly to the GSAB.

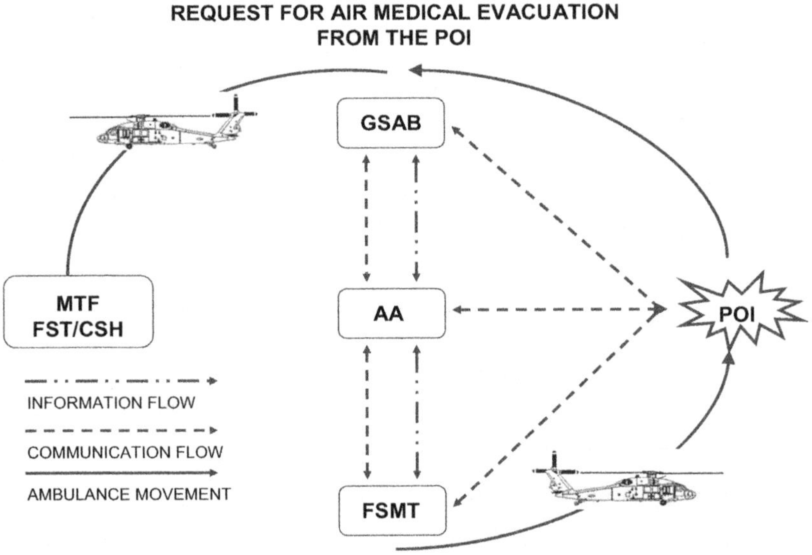

Figure 4-4. Air evacuation request in the corps

4-14. Figure 4-5 illustrates evacuation zones during a noncontiguous conflict. Here, no forward line of own troops (FLOT) is clearly visible.

- Forward support medical evacuation teams are placed where they are most needed. This can be with troops most often engaged in combat, high population density areas, area of famine or disease with high civilian casualties, refugee areas, geographically centralized location, and such.

- Area or general support FSMTs accomplish the patient transfer mission that develops between FSMTs and MTFs, between MTFs, and between MTFs and intertheater movement locations. These FSMT must also be responsible for the DS mission in their immediate vicinity.

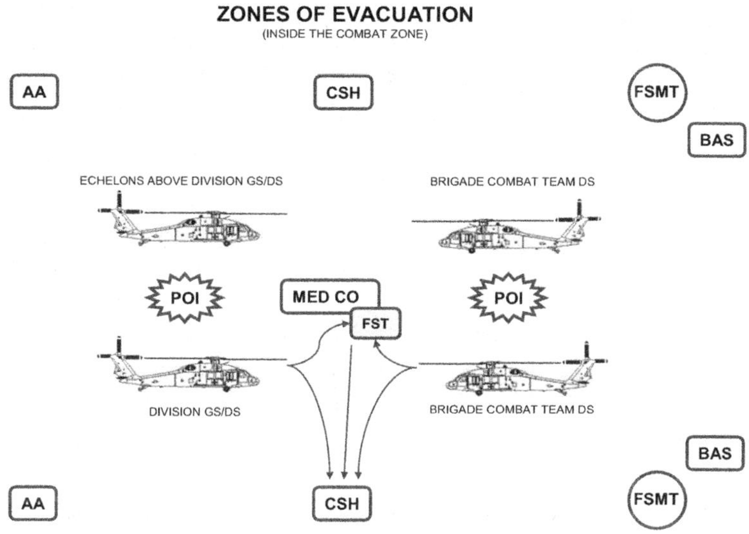

Figure 4-5. Air ambulance zones of evacuation

4-15. Figure 4-6 illustrates a comparison in aviation mission planning/execution cycle for medical evacuation and typical aviation.

- In contract to a typical aviation mission cycle, continuous aeromedical evacuation coverage results in extended operational duty days that often exceed 24 hours in length in order to provide seamless coverage to our forces. Medical evacuation units must plan and develop a detailed battle rhythm that addresses composite risk management, mission execution processes, and tactics, techniques, and procedures unique to 24-hour continuous evacuation operations.
- A medical evacuation crew cycle begins much like any other mission planning however they will not know the location or the time of execution for their mission. Since during these cycles it is critical the crew responds as rapidly as safely possible when the last information is received, it is imperative that crews are afforded the ability to manage rest while ensuring the procedures are in place to maintain battlefield situational awareness and facilitate a rapid, safe execution when a 9-line mission is received. This includes but is not limited to crew rest cycles, briefing procedures, and aviation and logistical support to remote/split-based crews to ensure the safe and timely execution of missions.

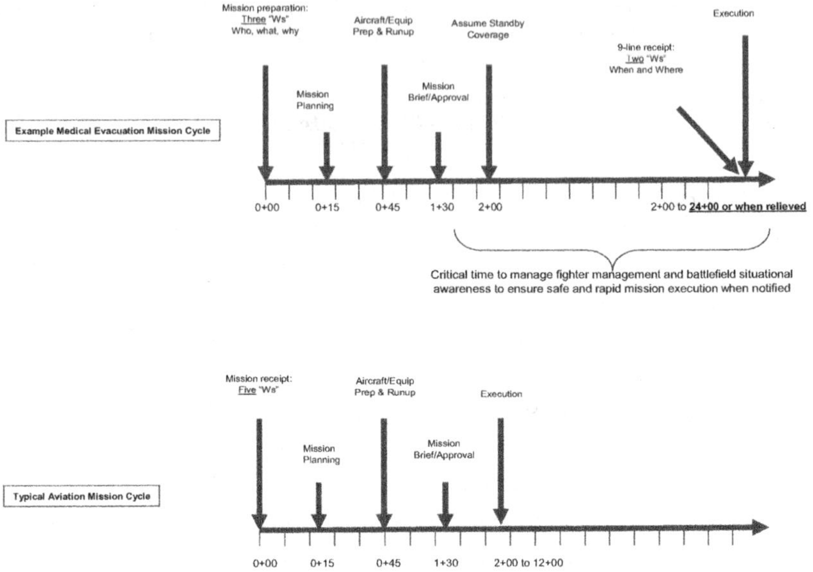

Figure 4-6. Aviation mission planning/execution cycle

JOINT INTERCONNECTIVITY

4-16. When directed by the GCC, Army evacuation assets may be used to evacuate patients from the other services or may be placed in direct support of other service, allied, or coalition units participating in the operation. Prior to the initiation of the operation, communications, and procedural issues should be synchronized to ensure the seamless execution of the operation. Interoperability issues can result in slower response times and may adversely effect medical evacuation operations. Figure 4-7 depicts the communications and patient flow in a joint scenario.

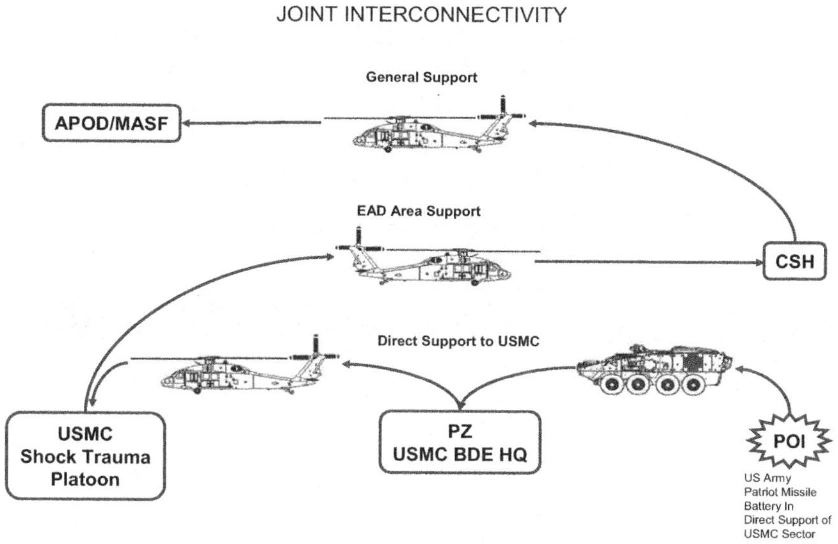

Figure 4-7. Air and ground ambulance evacuation in a joint environment

SECTION III — MEDICAL EVACUATION UNITS, ELEMENTS, AND PLATFORMS CONSIDERATIONS

MEDICAL EVACUATION SUPPORT PROTOCOL

4-17. Medical evacuation is performed by the higher role of medical care supporting forward and evacuating from the lower role. Medical evacuation support for theater, corps, and division is provided by organic ambulance teams, squads, and platoons. In addition, it is provided by the MMB medical company, ground ambulance medical company (GAMC), ASMC, and the GSAB medical company, AA.

4-18. The MMB maximizes the effective employment of ground ambulances resources whether in support of a division, corps, or at theater level. It coordinates through the combat aviation brigade's GSAB for effective use of AA support of its ASMCs. The medical brigade, MMBs, GSAB, sustainment brigade, BCTs and division surgeon section (DSS) coordinate and synchronize medical evacuation operations for the division AOs. The DSS coordinates with the MEDBDE for the evacuation of patients from division MTFs to supporting CSHs. The MEDBDE coordinates and arranges the required intratheater medical evacuation of patients to and from supporting CSHs. The MEDCOM coordinates for evacuation of patients out of the theater. To ensure that patients are evacuated to the appropriate MTFs, MROs are organic to the MEDCOM, and MEDBDE.

Medical Evacuation Platforms Factors

4-19. Evacuation platforms must be capable of keeping pace with the troops supported. The patient's medical condition is the overriding factor in determining the evacuation platform and destination MTF. The AA operates wherever needed on the battlefield, dependent on risk and METT-TC factors. Use of hardened armored MEV may be the vehicle of choice for some missions for short evacuation to an MTF or to a secure AXP for transfer to an air or wheeled ground ambulance. The criteria for an efficient and effective evacuation system were identified in Chapter 1.

4-20. The crew of the AA, assisted by onboard patient monitoring and diagnostic equipment, is trained in aeromedical procedures to provide optimum en route patient care. It is the platform of choice for most categories of patients. However, insufficient numbers of AAs are available to evacuate all patients expected in a battle zone. To conserve these valuable resources, medical planners should plan to use AAs to primarily move Priority I, URGENT, Priority IA, URGENT-SURG, and Priority II patients with other categories on a space-available basis.

SECTION IV — MEDICAL EVACUATION AT UNIT LEVEL

CONSIDERATIONS FOR MEDICAL EVACUATION MISSIONS

4-21. As METT-TC factors affect all medical evacuation missions and the employment of ambulance assets. The medical commander must consider the basic tenets that influence the employment of medical evacuation assets. These factors include the patient's medical condition and the:

- Tactical commander's plan for employment of combat forces.
- Enemy's most likely course of action.
- Anticipated patient load.
- Expected areas of patient density.
- Availability of medical evacuation resources to include ground and air crews.
- Availability, location, and type of supporting MTFs.
- Protection afforded medical personnel, patients, and medical units, vehicles, and aircraft under the provisions of the Geneva Conventions.
- Army airspace command and control plan.
- Engineer obstacle plans.
- Fire support plan (to ensure medical evacuation assets are not dispatched onto routes and at the times affected by the fire support mission).
- Road network/dedicated medical evacuation routes (contaminated and clean).
- Weather conditions.

Patient Acquisition

4-22. Units with organic medical evacuation assets have the primary responsibility for patient acquisition. Methods of employment and evacuation techniques differ depending upon the nature of the operation.

4-23. Units without organic ambulance assets are provided medical evacuation support on an area support basis. Units must develop techniques which facilitate the effective employment of their combat medics/company/platoon medics, enhance the ability to acquire patients in forward areas, and rapidly request medical evacuation support. The techniques developed should be included in the unit TSOP. As a minimum, the TSOP should include the—

- Vehicle assignment for the organic medical personnel.
- Vehicles designated to be used for casualty transport and/or patient evacuation.
- Procedures for requesting medical evacuation support (during routine operations or during mass casualty situations).
- Role of the first sergeant (1SG), platoon sergeants, and CLSs in medical evacuation.

Maneuver Battalion Medical Platoons

4-24. The medical platoon leader (a physician) along with the medical platoon field medical assistant should be included in all battalion tactical planning. This officer—

- Maintains knowledge of the concept of operations, commander's intent, and the anticipated medical requirements.
- Develops the HSS/FHP plan (FM 8-55 [FM 4-02.55] and FM 4-02.4) and provides HSS/FHP overlays with preplanned evacuation routes and CCPs, to ambulance squads or teams for inclusion in the battalion OPLAN.
- Ensures that the platoon sergeant provides strip maps or other navigational tools to the ambulance drivers, if needed.
- Requests augmentation support from the supporting medical company in advance of the operation, if required. When elements of other maneuver battalions are attached as part of a task force, then ensures that adequate medical elements are included in the support package.
- Ensures that orientation and support are provided for the medical personnel. This precludes taxing the medical elements of the receiving unit. These responsibilities are normally delegated to the medical operations officer (field medical assistant).

4-25. The ambulance/evacuation section squad leader/NCO ensures that the ambulance teams have a working knowledge of the terrain features in the AO. Whenever possible, he familiarizes with primary and secondary medical evacuation routes through route reconnaissance conducted by the squad leaders. This platoon sergeant and squad leader manages the employment of the ambulance teams and monitors the communications net to remain abreast of the tactical situation.

4-26. The following factors should be considered when selecting ambulance routes:

- Tactical mission.
- Coordinating evacuation plans and operations with the unit movement officer.
- Security of routes and security escort.
- Availability of routes.
- Physical characteristics of roads and cross-country routes (to include natural obstacles).
- Requirements to traverse roads in urban areas and potential obstructions from rubble and debris.
- Traffic density.
- Time and distance factors.
- Proximity of possible routes to areas that may be subject to enemy fire.
- Lines of patient drift.
- Cover, concealment, and available defilade for moving and stationary vehicles.
- Engineer obstacle plans.
- Fire support plan (to ensure medical evacuation assets are not dispatched onto routes and at the times affected by the fire support mission).

4-27. Depending upon the combat situation, the modes of evacuation may include walking soldiers who are wounded, manual and litter carries, nonmedical transportation assets, or dedicated medical evacuation platforms. Evacuation in the battalion area normally depends on the organic ambulances assigned. Evacuation by AA is dependent upon the patient's medical condition, availability of air assets, tactical situation, air superiority, and weather conditions.

4-28. The ambulance team is normally deployed to each maneuver company trains and one remains with the Role 1 MTF/BAS in the combat trains. Ambulances operate as far forward as the tactical situation permits. An ambulance team operating in a maneuver company AO, is normally under the tactical control of the maneuver company executive officer or 1SG. This ambulance team, however, remains under the OPCON of the medical platoon.

4-29. The medical operations officer ensures that the ambulances are located close to the anticipated patient workload. An ambulance team consists of one ambulance and two combat medics (on track vehicles, a third emergency care specialist is required to provide en route medical care). One or two of these teams serve in DS of a maneuver company. To become familiar with the specific terrain and battlefield situation, the team maintains contact with the supported maneuver company during most combat operations. Maneuver medical platoon ambulances not in DS are positioned strategically throughout the battalion area or are sited at the BAS to—

- Evacuate patients from the company aid posts, CCPs to the BAS.
- Reinforce the forward positioned ambulance teams.
- Support the combat forces held in reserve and/or scout, mortar and forward observer platoons.

4-30. Another employment option is to forward site the additional ambulance teams at company CCPs, as well as at the BASs. Many times the ambulance team finds battlefield casualties who have not been seen by a trauma specialist. In these cases, the team members dismount and then find, treat, and evacuate the patients. Ambulance teams not specifically dedicated to support combat elements can be used as messengers in medical channels and to provide transport of emergency medical personnel, equipment, and supplies.

4-31. During static situations where the maneuver company is not in enemy contact or is in reserve, the ambulance team returns to the BAS to serve as reinforcement to other elements in contact. However, during movement to contact, the ambulance team immediately deploys with its supported unit. In moving patients back to the CCP, the team may be assisted by nonmedical personnel. Specific duties of the ambulance team are to—

- Maintain contact with supported elements.
- Find and collect the wounded.
- Administer EMT.
- Initiate or complete the FMC.
- Evacuate patients to the BAS.
- Direct or guide ambulatory patients to the BAS.
- Resupply company/platoon medics.
- Serve as messengers in medical channels.
- Route recons from Role 1 MTF to Role 2 MTF or the higher role of medical care.

Casualty Collection Points

4-32. During the offense, CCPs may be used to avoid hampering the movement of the maneuver elements. In fast-moving situations, preplanned CCPs are included in the HSS plan and activated based on the crossing of phase lines, upon the occurrence of predetermined events, or on the execution of other control measures. It may be necessary to set up multiple CCPs for each phase of an operation. Rotating the use of these points precludes the enemy from using them to pinpoint maneuver elements or from attracting enemy fires. When the situation permits, medical evacuation of patients from CCPs, or AXPs may be accomplished by AAs.

4-33. Ambulance teams move using available terrain features for cover and concealment. They avoid prominent terrain features and likely targets. When stationary, the ambulance crew should conceal the vehicle as much as possible. When a casualty occurs in a tank, Bradley infantry fighting vehicle (BIFV), a Stryker wheeled armored vehicles, the ambulance team moves as close to the armored vehicle as possible. Assisted by the armored crew, if possible, the casualty is extracted from the vehicle and then administered EMT. The ambulance team moves the patient to the Role 1 MTF/BAS or to a CCP point to await further evacuation.

4-34. The senior company medic/trauma specialist normally remains with the company combat trains, but may be used anywhere in the company area, even assisting the ambulance teams in some situations. He may be used to direct ambulance teams to locations where vehicle crews need assistance or where injured or wounded crew members have been left. In some situations, crew members may have to rely on self-aid or buddy aid until the CLS or the platoon medic/trauma specialist arrives.

Area Medical Evacuation Support

4-35. Medical evacuation on an area basis is required at all roles in the AHS system. The division units without organic medical evacuation resources, such as combat engineers, will require evacuation support on an area basis. To ensure that these elements receive adequate support, the medical planner must include their requirements into the OPLAN. Prior coordination is essential to ensure that the locations of CCPs, AXPs, and BASs are disseminated to these elements and that any unique support requirements are included in the OPLAN. Further, the triggers for relocating these assets and their next locations should also be included.

MEDICAL EVACUATION SUPPORT FROM ROLE 1 AND ROLE 2 MEDICAL TREATMENT FACILITIES

4-36. Evacuation from Role 1 MTFs operated by treatment squad or team is normally provided by the BSMC/FSMC ambulance platoon and the GSAB AA company. Further, these ambulance assets provide medical evacuation support on an area basis to other units in the brigade area, as required.

Ground Ambulances

4-37. The ambulance teams from the ambulance platoon are normally collocated with the BSMC/FSMC treatment platoon for mutual support. They establish contact and locate one ambulance team with the medical platoon of each maneuver battalion. The remaining ambulances are used for BCT operations and area support. The ambulances are positioned at AXPs or CCPs, or are field-sited with Role 1 MTFs operated by the BSMC/FSMC treatment teams.

4-38. A FSMT from the supporting GSAB medical company, AAs provides DS AA support for the BSMC/FSMC Role 2 MTF. The teams normally remain under the operational control of the GSAB. The DS relationship provides authority to the BSMC/FSMC in coordination with the brigade support battalion/forward support battalion (BSB/FSB) distribution support section, brigade surgeons section to direct the integrated air and ground medical evacuation support for the BCT. The section leader of the FSMT should be included in the brigade tactical planning process.

Air Ambulances

4-39. The AA team evacuates Priority I, URGENT patients from as far forward as possible to the BSMC/FSMC. Further, when a FST is colocated with a BSMC/FSMC, AAs evacuate Priority IA, URGENT-SURG to this facility if the patient can not be flown to a Role 3 MTF condition due to distance or patient condition. External lift capabilities of aeromedical evacuation helicopters add an important dimension to its role on the battlefield in moving medical supplies and equipment.

4-40. The GSAB AAs may operate from the sustainment area and BSAs providing around-the-clock, immediate response, evacuation aircraft, as well as provide the BSMC/FSMC commander flexibility and agility in the movement of treatment teams and equipment to the forward battle area. It also provides the capability to rapidly resupply Class VIII supplies to combat units.

4-41. The medical elements within the BSB/FSB will request AA support through the brigade aviation element (BAE). The FSMT leader will provide coordination between the BSB/FSB, BAE, or the BCT S3.

4-42. In the BSMC/FSMC, the executive officer (XO) is the principal assistant to the commander for the tactical employment of the company assets. The XO should be included in all brigade tactical planning. Officer preparation will be required to reinforce or reconstitute forward HSS elements and to request augmentation through the brigade surgeon section (BSS), if required. The FSMT leader keeps the BSMC/FSMC XO apprised of the operational capability. This enables the XO to effect timely reinforcement or augmentation. The BSMC/FSMC XO must be familiar with the specific terrain and battlefield situation. Further, the XO should have a thorough understanding of the division and brigade commanders' ground tactical plan.

Brigade Medical Evacuation Plan

4-43. The BSS is responsible for developing the BCT medical evacuation plan. The BSMC/FSMC commander is responsible for the execution of brigade medical evacuation plan, to include the use of both ground and air assets. The BSS medical planner should include the medical company commander, XO, brigade S1, brigade XO, medical platoon leaders, FSMT leader, the BSB/FSB medical plans and operations officer, and the BAE company commander (if available) in the planning process.

Ground Ambulance Evacuation Brigade Role 2 Medical Treatment Facilities

4-44. Evacuation from the BSMC/FSMC is provided by ground ambulances from the GAMC or from AAs from the GSAB AA company. The GAMC provides support to the BSMCs and medical elements in the division AO. They usually evacuate only those patients who will not RTD within 72 hours.

Ground Ambulance Support for Echelons Above Brigade/Division Units

4-45. The ASMC ambulance platoon normally collocates with the treatment platoon for mutual support and area taskings. It performs ground evacuation and en route patient care for supported units in the sustainment area. The ASMC may also evacuate patients from the supported Role 2 MTFs in the sustainment area to the ASMC Role 2 MTF. The ambulance platoon is mobile in operations as its assets may be totally deployed at one time. The platoon normally forward stations a portion of its teams in support of those units in the sustainment area.

4-46. The remaining teams are used for GS support to division sustainment, movement and maneuver and fires units, and for reinforcing support or ambulance shuttles. Platoons or squads from the GAMC will be in DS, or OPCON to, and collocated with supported division units.

SECTION V — EXCHANGE OF PROPERTY

PROPERTY EXCHANGE AND PATIENT MOVEMENT ITEMS

4-47. United States Army medical evacuation operations require that whenever a patient is evacuated from one MTF to another, or is transferred from one ambulance to another, medical items of equipment (casualty evacuation bags [cold weather-type bags], blankets, litters, and splints) remain with the patient. To prevent rapid and unnecessary depletion of supplies and equipment, the receiving Army element exchanges like property with the transferring element. This reciprocal procedure will be practiced to the fullest extent possible through all phases of evacuation from the most forward element through the most rearward hospital.

4-48. With USAF aeromedical evacuation operations, a major factor in the evacuation of patients is that specific medical equipment and durable supplies designated as patient movement items (PMI) must be available to support the patient during the evacuation. Examples of PMI include—

- Ventilators.
- Litters.
- Patient monitors.
- Pulse oximeters.

These items will be available for exchange at the supporting ASFs and MASFs. Refer to JP 4-02 and FM 4-02.1 for additional information on PMI.

This subparagraph implements STANAG 2128 and QSTAG 436.

4-49. Medical property of allied nations (NATO and ABCA armies) accompanying patients of allied nations will be returned to the parent nation at once, if possible. If it is not possible, like items will be exchanged as in the paragraph above.

4-50. When medical property of coalition forces without ratified standardization agreements accompanies patients of coalition forces, it is returned as soon as practical. Commanders should consult with their staff judge advocate early in the planning process to ensure appropriate policy and procedures are developed and disseminated.

SECTION VI — MEDICAL EVACUATION SUPPORT FOR THE OFFENSE, THE DEFENSE, AND STABILITY OPERATIONS

MEDICAL EVACUATION SUPPORT FOR THE OFFENSE

4-51. The offense is the decisive form of war, the commander's only means of attaining a positive goal or of completely destroying an enemy force (FM 3-0). The offense is characterized by rapid movement, deep penetrations, aggressive action, and the ability to sustain momentum regardless of counterfires and countermeasures.

4-52. When considering the evacuation plans to support an offensive action, the medical planner must consider many factors (FM 8-55 [FM 4-02.55]). The forms of maneuver, as well as the enemy's capabilities, influence the character of the patient workload and its time and space distribution. The analysis of this workload determines the allocation of AHS resources and the location or relocation of MTFs. Evacuation support of offensive operations must be responsive to several essential characteristics. As operations achieve success, the areas of casualty density move away from the supporting facilities. This causes the routes of medical evacuation to lengthen. Heaviest patient workloads occur during disruption of enemy main defenses, at terrain or tactical barriers, during the assault on final objectives, and during enemy counterattacks. The accurate prediction of these workload points by the medical planner is essential if medical evacuation operations are to be successful.

4-53. In traditional combat operations, the major casualty area of the division BCTs is normally the zone of the main attack. As the main attack accomplishes the primary task of the division, it receives first priority in the allocation of combat power. The allocation of combat forces dictates roughly the areas which are likely to have the greatest casualty density. As a general rule, all BCT MTFs are located initially as far forward as combat operations permit. This allows the maximum use of these facilities before lengthening evacuation lines force their displacement forward. In operations that feature deep battles with CBRN weapons targeted at supporting logistical bases, MASCAL operations may be conducted in sustainment area.

4-54. As advancing combat formations extend control of the battle area forward, supporting medical elements overtake patients. This facilitates the acquisition of the battle wounded and reduces the vital time elapsed between wounding and treatment. In offensive operations, two basic problems confront the supporting evacuation units.

- First, contact with the supported unit must be maintained. Responsibility for the contact follows the normal HSS pattern of rear to front. The contact is maintained by forward deployed air and ground evacuation resources.
- Secondly, the mobility of the MTFs supporting the combat formations must be maintained. Periodically, BCT medical companies and collocated FSTs, are cleared of their patients so that they may move forward. In some scenarios, CSHs may also have to evacuate patients to reduce patient overflow, surgical backlogs, or to increase bed availability.

4-55. This requirement for prompt evacuation of patients from forward MTFs requires available ambulances to be leveled well forward from the outset. The requirement for periodic movement of large numbers of patients from divisional and corps facilities further stresses the evacuation system.

Types of Offensive Operations

Movement to Contact/Attack

4-56. Medical evacuation support in movement to contact/attack is key to the tactical plan. Prior deployment of evacuation resources with parent and supported units permits uninterrupted and effective evacuation support.

Exploitation and Pursuit

4-57. Evacuation support of exploitation and pursuit operations resembles those discussed for the envelopment (paragraph 4-59). Since exploitation and pursuit operations can rarely be planned in detail, evacuation operations must adhere to TSOPs and innovative C2. These actions are often characterized by—

- Decentralized operations.
- Unsecured ground evacuation routes.
- Exceptionally long distances for evacuation.
- Increased reliance on convoys and AAs.

Medical Evacuation Support for Different Forms of Offensive Maneuver

Penetration and Attack

4-58. In this tactic, the attack passes through the enemy's principal defensive position, ruptures it, and neutralizes or destroys the enemy forces. Of all forms of offensive maneuver, the penetration of main enemy defenses normally produces the heaviest medical evacuation workload. Patient acquisition starts slowly, but becomes more rapid as the attack progresses. The evacuation routes lengthen as the operation progresses. The penetration maneuver is often preceded by heavy preparatory fires which may evoke heavy return fire. These enemy fires may modify the decision to place evacuation assets as far forward as possible. Patient evacuation may be slow and difficult due to damage to roads or the inaccessibility of patients. Evacuation support problems multiply when some combat units remain near the point of original penetration. This is done to hold or widen the gap in enemy defenses while the bulk of division forces exploit or pursue the enemy. Treatment elements are placed near each shoulder of the penetration; ground evacuation cannot take place across an avenue of heavy combat traffic. Besides the heavy traffic, the area of the penetration is normally a target for both conventional and nonconventional CBRN weapons.

Envelopment

4-59. In the envelopment, the main or enveloping attack passes around or over the enemy's principal defensive positions. The purpose is to seize objectives which cut the enemy's escape routes and subject him to destruction in place from flank to rear. Since the envelopment maneuver involves no direct breach of the enemy's principal defensive positions, the medical evacuation system may not be confronted with a heavy workload in the opening phase—

- Ambulances are echeloned well forward in all levels of AHS to quickly evacuate the patients generated by suddenly occurring contact.
- Medical treatment facilities moving with their respective formations overtake patients during evacuation and reduce delays in treatment. After triage and treatment, the patients are evacuated to division/corps-level facilities by accompanying division/corps ground ambulance assets or AAs.

- When the isolated nature of the envelopment maneuver precludes prompt evacuation, the patients are carried forward with the treatment element. Again, nonmedical vehicles may be pressed into emergency use for this purpose. When possible, nonmedical vehicles should be staffed by a combat medic or CLS.
- When patients must be carried forward with the enveloping forces, AHS commanders use halts at assembly areas and phase lines to arrange combat protection for ground ambulance convoys through unsecured areas.
- Further, the commander may take advantage of friendly fires and suppression of enemy air defenses to call for prearranged AA support missions, or emergency use of medium-lift helicopter backhaul capabilities.

Infiltration

4-60. Infiltration is a form of maneuver used during offensive operations.

- The division can attack after infiltration or use it as a means of obtaining intelligence and harassing the enemy.
- Though it is not restricted to small units or dismounted actions, the division employs these techniques with a portion of its units, in conjunction with offensive operations conducted by the remainder of its units.

Army Health System of Infiltration/Maneuver

4-61. The AHS of infiltration/maneuver is restricted by the amount of medical equipment, supplies, and transportation assets that can be introduced into the attack area. No deployment of division-level medical units without their organic transportation should be attempted.

4-62. Elements of unit-level AHS should be accompanied by their organic vehicles, and ambulances should receive priority for deployment. It may be necessary to man-carry enough BAS equipment into the attack area to provide EMT and ATM; however, this results in degrading mobility. When the element is committed without its ambulances, patients are evacuated to the BAS by litter bearer teams. This requires reinforcement of the medical platoon by division or corps medical personnel or improvisation of litter teams using combat troops (if available and approved by the tactical commander). Patient evacuation from the BAS and medical resupply of the force may be provided by litter bearers, depending upon distances and degree of secrecy required.

4-63. When airborne and air assault forces are used, infiltrating elements may land at various points within the enemy's rear area and proceed on foot to designated attack positions. As in surface movement, the amount of medical equipment taken may be limited. In airborne operations, the evacuation of patients will be by litter bearers or frontline ambulances to CCPs or the BAS and then by division-level ambulances to the Role 2 MTF. In air assault operations, the evacuation is by litter bearers to CCPs or to the BAS and then by AAs to a Role 2 MTF. Once the combat element begins the assault on the objective, secrecy is no longer important and its isolated location requires HSS characteristic to airborne and air assault operations until ground linkup.

Turning Movement

4-64. The turning movement is a variant to the envelopment in which the attacker attempts to avoid the defense entirely; rather, the attacker seeks to secure key terrain deep in the enemy's rear and along his LOC. Faced with a major threat to his rear, the enemy is thus "turned" out of his defensive positions and forced to attack rearward at a disadvantage.

Note. General MacArthur's invasion at Inchon during the Korean War is an example of a classic turning movement. Casualties were initially light as the main defenses were avoided; however, as the invasion developed, resistance stiffened and higher casualty rates were experienced. Further, as fighting occurred in a populated area (Seoul), significant civilian casualties resulted. The lack of Korean health care providers caused many of these civilians to seek medical aid from US field medical units.

Medical Evacuation for Turning Movement

4-65. Medical evacuation support to the turning movement is provided basically in the same manner as to the envelopment. As the operation is conducted in the enemy's rear area, LOC and evacuation routes may be unsecured resulting in delays in resupply and evacuation. In the Inchon example, a hospital ship was located off the coast to accept patients evacuated from the fighting. However, due to the precarious tides, evacuation and resupply were often delayed for hours and sometimes days since the harbor could not be navigated by small vessels. It was not until Kimpo Airfield fell that timely evacuation could occur. The deployed HSS units must be able to quickly clear the battlefield of patients, evacuate them from the forward areas, and sustain the patients in rear areas until evacuation routes are established.

MEDICAL EVACUATION SUPPORT FOR THE DEFENSE

TYPES OF DEFENSE

4-66. There are three forms of the defense:

- **Area defense** concentrates on denying enemy access to designated terrain for a specific period of time, rather than on the outright destruction of the enemy.
- **Mobile defense** focuses on denying the enemy force by allowing him to advance to a point where he is exposed to a decisive counterattack by the striking force. The primary defeat mechanism, the counterattack, is supplemented by the fires of the fixing force.
- **Retrograde** is an organized movement to the rear and away from the enemy. The enemy may force these operations or a commander may execute them voluntarily. Within the retrograde operation there are three forms: delay, withdrawal, and retirement.

4-67. Support is generally more difficult to provide in the defense. The patient load reflects lower casualty rates, but forward area patient acquisition is complicated by enemy actions and the maneuver of combat forces.

4-68. Medical personnel generally are permitted much less time to reach the patient, complete vital EMT, and remove him from the battle site. Increased casualties among exposed medical personnel further reduce the medical treatment and evacuation capabilities. Heaviest patient workloads, including those produced by enemy artillery and CBRN weapons, may be expected during the preparation or initial phase of the enemy attack and in the counterattack phase.

4-69. The enemy attack may disrupt ground and air routes and delay evacuation of patients to and from treatment elements. The depth and dispersion of the defense create significant time and distance problems for evacuation assets. Combat elements may be forced to withdraw while carrying their remaining patients to the rear. The enemy exercises the initiative early in the operation which may preclude accurate prediction of initial areas of casualty density. This makes the effective integration of air assets into the evacuation plan essential. The use of AAs must be coordinated and synchronized with the supporting GSAB to ensure the synchronized execution of evacuation operations occurs.

4-70. The support requirements for retrogrades may vary widely depending upon the tactical plan, the enemy reaction, and the METT-TC factors. Firm rules that apply equally to all types of retrograde operations are not feasible, but considerations include:

- Requirement for maximum security and secrecy in movement.

- Influence of refugee movement that may impede medical evacuation missions conducted in friendly territory.
- Integration of evacuation routes and obstacle plans should be accomplished.
- Difficulties in controlling and coordinating movements of the force which may produce lucrative targets for the enemy.
- Movements at night or during periods of limited visibility.
- Time and means available to remove patients from the battlefield. In stable situations and in the advance, time is important only as it affects the physical well-being of the wounded. In retrograde operations, time is more important. As available time decreases, medical managers at all levels closely evaluate the capability to collect, treat, and evacuate all patients.
- Medical evacuation routes will also be required for the movement of troops and materiel. This causes patient evacuation in retrograde movements to be more difficult than in any other type of operation.
  - Command, control, and communications may be disrupted by the enemy.
  - Successful medical evacuation requires including ambulances on the priority list for movement.
  - Providing for the transportation of the slightly wounded in cargo vehicles and providing guidance to subordinate commanders defining their responsibilities in collecting and evacuating patients.
  - Special emphasis must be placed on the triage of patients and consideration given to the type of transportation assets available for evacuation.
- When the patient load exceeds the means to move them, the tactical commander must make the decision as to whether patients are to be left behind. The medical staff officer keeps the tactical commander informed in order that he may make a timely decision. Medical personnel and supplies must be left with patients who cannot be evacuated (refer to FM 4-02 for additional information).

Air Ambulance Evacuation Support to Covering Forces in Defensive Operations

4-71. The division covering force uses combat power to screen the FLOT and force the early deployment of the enemy's main force. Organic AHS elements of the covering forces establish aid stations of minimal size or operate in the split-team mode (FM 4-02.4). Ambulances are deployed well forward to promptly clear patients from combat units. Medical care is limited to EMT and ATM, followed by rapid evacuation whenever possible.

4-72. The covering force surgeon maintains communications with attached aviation elements and uses these assets (augmented by medical personnel to provide en route medical care when feasible) to provide backhaul of casualties, rather than use ground ambulances. The early evacuation of patients from BASs ensures their mobility for rearward displacement. The BCT or armor cavalry regiment (ACR) medical company/troop is equipped to receive patients generated in the covering force area. Depending upon the fluidity of the defensive situation, the FSMT (if available) may have to rapidly evacuate patient's further rearward to a BSMC/FSMC and collocated FST or a CSH.

4-73. The wide dispersion of units and the manner in which they withdraw make patient acquisition difficult. When covering forces withdraw, patients are transported to the rear by the medical element that acquired them. Seriously wounded or injured patients are given priority for evacuation by air. There are usually few COSRs during covering force operations, although delayed symptoms may occur once the element is withdrawn to a safe area. Ground ambulances augmented by nonmedical transportation assets evacuate the remaining patients.

MEDICAL EVACUATION SUPPORT FOR STABILITY OPERATIONS

4-74. Stability operations promote and protect US national interests by influencing the threat, political, and information dimensions of the operational environment. They include developmental, cooperative activities during peacetime and coercive actions in response to crisis.

4-75. Some of the purposes for which the US Forces are employed to conduct stability operations are as follows:

- Civil control.
- Civil security.
- Provide essential services.
- Support economic and infrastructure development.
- Governance.

4-76. Stability operations may complement the main effort or they may themselves constitute the main effort. Table 4-2 identifies the types of stability operations and the types of missions the BCT may conduct.

Table 4-2. Types of stability operations

Types Of Operations	*Types Of Missions*
Civil Control	Establish police forces training.
	Integrate trained police into operations.
	Counter organized crime.
	Establish judicial system.
	Transition to indigenous police forces.
Civil Security	Initiate security forces training.
	Establish National Security Forces.
	Integrate security forces into operations.
	Isolate population from insurgency.
	Defeat insurgency.
	Transition to National Security Forces.
Essential Services	Sewage treatment plants operating.
	Establish trash disposal.
	Water treatment plants operating.

Table 4-2. Types of stability operations

	Restore electrical power. Reopen hospitals and clinics.
	Reopen schools and universities.
Economic/Infrastructure Development	Implement Employment Programs.
	Secure vital natural resources.
	Repair/rebuild distribution infrastructure.
	Prioritize reconstruction projects.
	Essential banking services available.
	Implement public works projects.
Governance	Identify and recruit local leaders.
	Facilitate establishment of sector representation.
	Facilitate establishment of neighborhood councils.
	Facilitate establishment of district councils.
	Support and secure elections.

4-77. Medical evacuation support to forces deployed in stability operations is dependent upon the specific type of operation, anticipated duration of the operation, number of forces deployed, theater evacuation policy, medical troop ceiling, and anticipated level of violence. In most situations, medical evacuation support follows the traditional support provided to combat forces. If there is a shortened theater evacuation policy, a limited medical troop ceiling, and limited hospitalization assets within the AO, organic and DS ambulance support is provided from the POI to the supporting Roles 1 or 2 MTF and, once the patient is stabilized for further evacuation, from the treatment element to an airfield for evacuation out of the theater.

4-78. Reconstruction operations use Army forces to assist civil authorities, foreign or domestic, as they prepare for or respond to crises and relieve suffering. In reconstruction operations, Army forces provide essential support, services, assets, or specialized resources to help civil authorities deal with situations beyond their capabilities. The purpose of reconstruction operations is to meet the immediate needs of designated groups for a limited time, until civil authorities can do without Army assistance.

4-79. Humanitarian and civic assistance operations can include a number of activities such as disaster relief, civil support, refugee assistance, the provision of medical care to isolated populations, and refeeding programs resulting from famines or natural disasters. Medical evacuation assets may be used to evacuate the injured from disaster sites, to provide the emergency transport of critically needed medical supplies and personnel to remote locations, or to perform emergency rescues during times of flooding, wildfires, or other natural disasters.

SECTION VII — MEDICAL EVACUATION SUPPORT FOR ENABLING/SHAPING OPERATIONS

PASSAGE OF LINES

4-80. Passage of lines presents a challenge for the medical planner. There will be a number of medical evacuation units using the same air and road networks. Coordination and synchronization are essential if confusion and over evacuation are to be avoided. The information required to operate in the BCT or division AO includes:

- Radio frequencies and call signs.
- Operations plans and TSOPs.
- Location of MTFs.
- Location of CCPs and AXPs.
- Main supply route, forward arming and refueling points (FARPs), and A2C2 data.

SECURITY OPERATIONS

4-81. In security operations, the covering forces are dependent upon organic resources found in the maneuver battalion medical platoon for initial support. The level of command for the covering force (division or corps) determines the responsibility for the subsequent evacuation plan. In a corps covering force, for example, the corps AHS structure has the responsibility for establishing and operating the medical evacuation system to support the forward deployed corps forces. This is done to prevent the divisions following the covering forces from becoming overloaded with patients prior to the hand off and passage of lines. The use of CCPs, AXPs, and nonmedical transportation assets (CASEVAC) to move the wounded is essential. The covering force battle may be extremely violent. Patient loads will be high and the distance to MTFs may be much longer than usual. The effectiveness of the medical evacuation system depends upon the forward positioning of a number of ground ambulances and the effective integration of AAs into the evacuation plan.

ADVANCE, FLANK, AND REAR GUARDS

4-82. In advance, flank, and rear guards, these forces normally receive medical evacuation support through the attachment of evacuation teams. The teams evacuate patients to predesignated CCPs along a main axis of advance or to the nearest treatment element providing area support. Employment of AAs provides a measure of agility and flexibility.

RIVER CROSSING OPERATIONS

4-83. The river barrier itself exerts decisive influence on the use of division and BCT medical units. Attack across a river line creates an AHS delivery problem comparable to that of the amphibious assault. Army health system protection elements cross as soon as combat operations permit.

4-84. Early crossing of treatment elements reduces turnaround time for all crossing equipment that is used to load patients on the far shore. Maximum use of AA assets is made to prevent excessive patient buildup in far shore treatment facilities. Near shore MTFs are placed as far forward as assault operations and protective considerations permit to reduce ambulance shuttle distances from off-loading points. For detailed information on river crossing operations, refer to FM 90-13.

RECONNAISSANCE OPERATIONS

4-85. The reconnaissance in force is an attack to discover and test the enemy's position and strength or to develop other intelligence. The division usually probes with multiple combat units of limited size, retaining sufficient reserves to quickly exploit known enemy weaknesses.

4-86. Ambulances are positioned well forward at both unit and division levels. Ambulances are moved at night to enhance secrecy. The echeloning of ambulances is an indication to the enemy that an attack is imminent due to the forward placement of HSS. Role 2 MTFs are not established until a significant patient workload develops.

4-87. Patients received at BASs of reconnoitering units are evacuated to Role 2 MTFs as early as practical or are carried forward with the force until a suitable opportunity for evacuation presents itself. Maximum possible use of AA assets is made to cover extended distances and to overcome potentially unsecured ground evacuation routes.

RETROGRADE OPERATIONS

4-88. A retrograde operation is a movement to the rear or away from the enemy. This type of operation may be forced by enemy action or may be executed voluntarily.

AEROMEDICAL EVACUATION SUPPORT IN RETROGRADE OPERATIONS

4-89. Aeromedical evacuation support in retrograde operations may vary widely depending upon the enemy reaction and the situation. Firm rules that apply equally to all types of retrograde operations are impossible to establish. Factors to consider in planning aeromedical evacuation support for retrograde operations include—

- Patient's condition/status.
- Mission, enemy, terrain and weather, troops and support available, time available and civil considerations.
- Requirement for maximum efforts in secrecy and employment of aircraft survivability equipment.
- Influence of refugee movement (which may impede military medical movements conducted in friendly territory).
- Integration of evacuation routes and other Army aviation routes.
- Difficulty in controlling and coordinating movements of the force which may produce lucrative targets for the enemy.

4-90. During retrograde operations, battalion, brigade, and division surgeons must evaluate their capability to collect, treat, and evacuate patients. Three factors influence the number of patients evacuated from the battlefield and the time and place of patient treatment:

- Time available prior to MTF displacement.
- Location of the enemy advance.
- Means of evacuation.

4-91. Due to congestion on main supply routes and general evacuation routes, the AAs may provide the best means of evacuation during retrograde operations. Seriously wounded patients should be evacuated by the fastest and most comfortable means and should receive medical care en route. Proper sorting (triage) and rapid evacuation of patients lessen the need for establishing a complete Role 2 MTF operation. Therefore, the FSMT (if available) leader should be prepared to surge his efforts in the retrograde. Combat and operational stress reaction casualty rates are usually low as compared to wounded in action (WIA) rates during controlled retrograde operations, but increase after safety is reached.

4-92. Commanders must decide to displace medical evacuation units before they hamper the maneuver forces conducting the retrograde movement. During mission planning, the medical evacuation units/ elements integrate with the supported unit's displacement plan. The medical evacuation element leaders must coordinate with the supported unit commander and the BSMC for displacement decision points and locations.

WITHDRAWAL OPERATIONS

4-93. A withdrawal operation is one in which a deployed force disengages from an enemy force. It may be forced by enemy pressure or conducted voluntarily.

4-94. Although the deployed force disengages from the enemy, contact is maintained by security elements while the main force moves to the sustainment area, forms march columns, and moves to a predesignated location. Aeromedical evacuation elements must be prepared to support the withdrawing force as well as any forces left in contact. Time available to acquire, treat, and evacuate patients from the line of contact is limited.

AEROMEDICAL EVACUATION SUPPORT IN WITHDRAWAL OPERATIONS

4-95. The AA medical company element supporting the withdrawing force must ensure it maintains sufficient station capability at its present location to support the patients while taking action to move to a new location.

4-96. In addition to the evacuation support provided to the force left in contact, there is a requirement to support the force moving to the sustainment area. If the main force infiltrates to the sustainment area, patients are carried by their parent units to the assembly area or MTFs in the sustainment areas. Medical assets are positioned in the assembly area to receive patients.

4-97. The division establishes Role 2 MTFs at a minimal size and well to the rear of the first line of alternate positions. If combat and environmental conditions indicate a light patient workload and the road network permits rapid ambulance movement, the division may use another Role 2 MTF to provide GS to all withdrawing brigades.

4-98. Role 2 MTFs conduct leapfrog rearward, occupying successive positions placed along the withdrawal route to minimize the requirement for multiple displacements by any one MTF. This avoids unnecessary interference with combat operations while providing continuous medical support. With the exception of an ambulance squad to support the covering force the division normally does not employ ambulances further forward than the BAS. AA elements will normally evacuate to sustainment area and corps positions to avoid impeding the withdrawal of brigades.

4-99. Preparation for the withdrawal operation includes distribution of extra medical consumable supplies and nonexpendable exchange items to each medical element. This is done to overcome the potential effects of isolated MTFs or the possible intermittent operation of the medical evacuation system. The AA medical company may receive the mission to assist with this distribution.

4-100. If withdrawal is made under enemy pressure, the provision of aeromedical evacuation is modified. Since the time available is critical, aeromedical resources cannot remain in the forward areas. The assets are usually withdrawn as a unit. Patients occurring during the withdrawal are carried to the sustainment areas by the parent unit, normally using nonmedical vehicles.

DELAYING OPERATIONS

4-101. Delaying operations occur when forces are insufficient to attack or to defend and when the defensive plan calls for drawing the attacker into an unfavorable situation. The delay maneuver may include friendly maneuver to successive positions across a broad front.

4-102. Delaying brigades split their combat power, moving their less mobile forces directly to the next defensive position while the elements remaining in contact fight to the sustainment area.

4-103. Aeromedical support for a delaying operation is conducted like withdrawal operations.

RETIREMENT OPERATIONS

4-104. A retirement operation is a rearward movement of a force not in contact with the enemy. It is conducted according to the force's OPLAN and without pressure by enemy forces.

4-105. Because the division is no longer in contact with the enemy, it can march (in multiple columns) directly to the rear. The AHS requirements for this type of operation are similar to those in a movement to contact. The patient workload is light. The FSMT (if available) leader will coordinate with the BSMC commander, and the AA medical company CP will coordinate with the DSS for guidance in the positioning of company assets to support this operation.

SECTION VIII — SPECIAL FORCES OPERATIONS

4-106. Army special operations forces (ARSOF) have limited organic AHS resources. They are, therefore, dependent upon the theater AHS for the majority of their health care requirements. To ensure that ARSOF receive comprehensive and timely AHS and logistics, medical planners must consider each medical function and determine the support organic and theater AHS assets will provide. The significant challenge for ARSOF medical planners is to ensure the efficient delivery and use of the conventional Class VIII (medical materiel) system for support of ARSOF missions.

4-107. Army special operations forces consist of special forces (SF), rangers, special operations aviation (SOA), psychological operations (PSYOP), and CA. The SOA, CA, and PSYOP have limited medical capabilities and are very dependent upon the theater HSS assets and the organizations they are supporting. Special forces and rangers often operate far removed from conventional AHS and must be more self-reliant and sustaining than conventional forces. Accordingly, SF and ranger medical personnel receive enhanced medical training above that provided for a trauma specialist. The Army educates the SOF medic to be an independent care practitioner qualified to provide ATM to combat casualties. When deployed on independent operations, the two SF medics are the sole source of medical care for their operational detachment, the indigenous forces, and their families that the detachment supports. They can train the indigenous populace in basic medical skills and establish an austere AHS system. Nonmedical ARSOF personnel receive medical training at the combat lifesaver level. Special operation combat medics (SOCMs) are trained in emergency medical treatment and in combat trauma management, but they do not possess the invasive surgical skills of the SF medical sergeant. The SOCMs provide medical care in combat environments where resuscitative surgery is more rapidly accessible than in the small unit missions of SF. Physicians and physician assistants assigned to units with SOCMs can provide ATM. Army health system planning for missions where the risk of penetrating trauma is high and further complicated by time and extended distance requires extensive planning and coordination. The capability for medical personnel and units that can perform resuscitative surgery and invasive stabilization must be provided regardless of the ARSOF unit involved.

4-108. Although the ARSOF health care provider receives enhanced medical training exceeding the level and scope found in conventional forces, he depends heavily on the conventional AHS system to conserve the combat strength of the ARSOF (particularly in the area of medical evacuation where the ARSOF does not have a dedicated system).

4-109. Ideally, medical evacuation for ARSOF personnel should follow the doctrinal flow sequence. The ARSOF medical planner must be innovative and follow the tenets of immediate far forward stabilization. He directs evacuation to the appropriate MTF when the condition of the patient warrants it, with whatever means of transportation are available.

4-110. Medical evacuation of ARSOF casualties is an operational matter. That is, it must reflect the commander's concept of the operation. It can only succeed when the medical planner integrates the medical evacuation plan with the tactical plan and logistics airflow.

4-111. Conventional medical evacuation assets are not normally used when operations occur in hostile and denied areas. Use the conventional medical evacuation system once the casualty is extracted from the hostile or denied area. For additional information on AHS to ARSOF, refer to FM 3-05, FM 4-02.6. FM 8-42 (FM 4-02.42), and FM 4-02.43.

SECTION IX — URBAN OPERATIONS

TERRAIN AND ENVIRONMENT

4-112. Throughout history, battles have been fought on urbanized terrain. Some recent examples urban operations (UO) include Hue, Beirut, Panama City, and Baghdad. Military operations on urbanized terrain are those military actions planned and conducted on a terrain where man-made structures impact on the tactical options available to the commander. This terrain is characterized by a three-dimensional battlefield, having considerable rubble, ready-made fortified fighting positions, and an isolating effect on all movement and maneuvers and fires, and sustainment units. In this environment, the requirement for a sound and understandable evacuation plan cannot be overstated. Of concern to medical and tactical planners is the need to plan, train, prepare, and equip for evacuation from under, above, and at ground level.

TERRAIN CONSIDERATIONS

4-113. Medical evacuation in the urban environment is a labor-intensive effort. Due to rubble, debris, barricades, and destroyed roadways, much of the evacuation effort will be by manual litter teams. When this occurs, establish a litter or ambulance shuttle system. The shuttle system reduces the distance that litter teams have to carry the wounded or injured Soldiers. This enhances the litter teams' effectiveness by providing brief respites, reducing fatigue. Further, the litter teams stay in the forward areas. They are familiar with the geography of the area and what areas have or have not been searched for casualties. In moving patients by litter, you should:

- Use covered evacuation routes such as subways, whenever possible.
- Use easily identifiable points for navigation CCPs.
- Rest frequently by alternating litter teams.

EVACUATION PLANNING CONSIDERATIONS FOR URBAN OPERATIONS

4-114. Conducting medical evacuation operations in the urban environment challenges the medical planner. He must ensure that the HSS/FHP plan includes special or unique materiel requirements or improvised use of standard equipment. The plan must be sufficiently flexible to support unanticipated situations.

4-115. Special equipment requirements include, but are not be limited to:

- Axes, crowbars, and other tools used to break through barriers.
- Special harnesses; portable block and tackle equipment; grappling hooks; collapsible litters; lightweight collapsible ladders; heavy gloves; and casualty blankets with shielding. This equipment, using pulleys, is for lowering casualties from buildings or moving them from one building to another at some distance above the ground.
- Equipment for the safe and quick retrieval from craters, basements, sewers, and subways. Casualties may have to be extracted from beneath rubble and debris.

4-116. Air ambulances equipped with a rescue hoist may be able to evacuate casualties from the roofs of buildings or may be able to insert needed medical personnel and supplies.

COMMUNICATIONS IN THE URBAN ENVIRONMENT

4-117. The urban environment degrades communications. Line of sight radios will be ineffective and individual Soldiers will not have access to radio equipment. The medical evacuation teams will have difficulty in locating injured or wounded Soldiers due to their isolation within buildings, or by being hidden in rubble and debris. Injured Soldiers can display alternate forms of communications, such as markers, panels, or field expedients (field/uniform jackets or T-shirts) indicating where they may be found.

Establishing Casualty Collection Points

4-118. Preplan and establish CCPs at relatively secure areas accessible to both ground and AAs. The location of these points are depicted on the HSS/FHP overlay to the OPLAN. Casualty collection points should—

- Offer cover from enemy fires.
- Be as far forward as the tactical situation permits.
- Be identifiable by an unmistakable feature (natural or man-made).
- Allow rapid turnaround of ambulances.

4-119. The tactical commander must approve route markings to the MTF and if to display of the Geneva Red Cross at the facility. (Camouflaging or not displaying the Geneva Red Cross can forfeit the protections, for both medical personnel and their patients, afforded under the Geneva Conventions. Refer to Appendix A and FM 4-02 for additional information.) The location of the MTF must be as accessible as possible, but well separated from fuel and ammunition depots, motor pools, reserve forces, or other lucrative enemy targets, as well as civilian hazards such as gas stations or chemical factories.

Ground Evacuation For Urban Operations

4-120. When using ground evacuation assets in support of UO, the medical planner must be aware that built-up areas may have significant obstacles to vehicular movement. Factors requiring consideration include the following:

- Rubble and other battle damage complicate and canalize transportation operations within the urban terrain.
- Bypassed pockets of resistance and ambushes pose a constant threat along evacuation routes.
- Land navigation with most tactical maps proves to be difficult. Using commercial city maps when available can aid in establishing evacuation routes.
- Ambulance teams must dismount from the ambulance, search for, and rescue casualties.
- Movement of patients becomes a personnel intensive effort. There are insufficient medical personnel to search for, collect, and treat the wounded. Medical units may require assistance in the form of litter bearers and search teams from supported units.
- Refugees may hamper movement into and around urban areas.
- Dislocated civilians, EPW, and detainees receive medical treatment according to the command policy and the Geneva Conventions.

Air Evacuation for Urban Operations

4-121. When using AA assets in support of UO, the medical planner must consider enemy air defense capabilities and terrain features, both natural and man-made, within and adjacent to the built-up areas. Air ambulances are the preferred means of evacuation in UO. Considerations in the use of AAs include the following:

- Movement is highly restricted and is canalized over secured areas, down wide roads, and open areas.
- Telephone and electrical wire and communication antennas hinder aircraft movement.
- Secure landing zones (LZs) must be available.
- Landing zones may include buildings with helipads on their roofs or sturdy buildings, such as parking garages.
- Snipers with air defense capabilities may occupy upper stories of the urban area's taller buildings.

SPECIALTY TRAINING

4-122. Medical personnel require special training in the tactics, techniques, and procedures required to operate in an UO environment. If they are to survive in this environment, they must know how to:

- Cross open areas safely.
- Avoid barricades and mines.
- Enter and depart safely from buildings.
- Recognize situations where booby traps or ambushes are likely and would be advantageous to the enemy. Detailed information on the conduct of combat operations in the urban environment is contained in FM 3-06 and FM 3-06.11.

Note. Medical personnel do not engage in offensive-type actions. They must rely on the supported unit to provide covering fires and to clear rooms and buildings prior to entry.

4-123. Units can modify and apply many of the extraction and patient evacuation techniques used in a mountainous terrain for evacuation in UO.

SECTION X — HOMELAND SECURITY

4-124. To preserve the freedoms guaranteed by the Constitution of the US, the Nation must have a homeland that is secure from threats and violence, including terrorism. Homeland security (HS) is the Nation's first priority, and it requires a national effort; in which the DOD has a key role in that effort. The National Strategy for Homeland Security (NSHS) complements the National Security Strategy of the US by providing a comprehensive framework for organizing the efforts of federal, state, local, and private organizations whose primary functions are often unrelated to national security. Critical to understanding the overall relationship is an understanding of the distinction between the role that DOD plays with respect to securing the Nation and HS, and the policy in the NSHS, which has the Department of Homeland Security (DHS) as the lead. Homeland security at the national level has a specific focus on terrorist threats. The DOD focus in supporting HS is broader. Military application of the NSHS calls for preparation, detection, deterrence, prevention, defending, and responding to threats and aggression aimed at the homeland. The DOD also provides military assistance to civil authorities (MACA), including consequence management activities. The Armed Forces of the US support the NSHS through two distinct but interrelated mission areas–homeland defense (HD) and civil support. The United States Northern Command (USNORTHCOM) is the unified command that has the DOD responsibility for HD and civil support.

HOMELAND DEFENSE

4-125. Homeland defense is the protection of US territory, sovereignty, domestic population, and critical infrastructure against external threats and aggression. The purpose of HD is to protect against and mitigate the impact of incursions or attacks on sovereign territory, the domestic population, and defense critical infrastructure. For HD missions, as directed by the President of the US and/or the Secretary of Defense, DOD serves as the lead federal agency.

HOMELAND DEFENSE CHARACTERISTICS

4-126. The DOD is the lead, supported by other agencies, in defending against traditional external threats/aggression (such as air and missile attack). However, against internal asymmetric, nontraditional threats (such as terrorism), DOD may be in support of DHS. When ordered to conduct HD operations within US territory, DOD will coordinate closely with other federal agencies or departments.

TYPES OF HOMELAND DEFENSE OPERATIONS

4-127. The types of homeland defense operations that the future brigade combat teams is expected to perform include—

- Offensive operations.
- Defensive operations.
- Humanitarian and civic assistance.
- Support to counterdrug operations.
- Show of force.
- Domestic support operations.
- Security operations (screen, guard, area security, and local security).
- Reconnaissance operations (zone, route, and area reconnaissance).

CIVIL SUPPORT

4-128. Civil support missions consist of military support to US civil authorities for domestic emergencies, and for designated law enforcement within the scope of restrictions required by the Posse Comitatus Act and other support approved by the Secretary of Defense (see FM 3-07).

4-129. Employment of military forces within the US, its territories, and possessions, under the auspices of civil support, typically falls under the broad mission of MACA. The MACA missions consist of three mission subsets. These mission subsets consist of military support to civil authorities, military support to civilian law enforcement agencies, and military assistance for civil disturbances.

MILITARY SUPPORT TO CIVIL AUTHORITIES

4-130. Military support to civil authorities (MSCA) is the most widely recognized form of DOD civil support because it usually consists of support for high-profile emergencies such as natural or man-made disasters that often invoke Presidential or state emergency/disaster declarations:

- **Natural Disasters.** In the event of a natural disaster or emergency, there will be a concerted US Government effort to support the affected areas. The DOD may be asked to provide assistance to DHS/Federal Emergency Management Agency (FEMA) in an attempt to save lives, protect property, and lessen the threat of a catastrophe in the US. Examples of natural disasters include, but are not limited to severe weather, wildland fire fighting, and animal disease eradication. When natural disasters occur and military assistance is anticipated, DHS/FEMA will request a defense coordinating officer (DCO) who serves as the single DOD point of contact within the disaster area. The DCO will be under the OPCON of the designated supported combatant command. Units supporting the event are normally OPCON to the supported combatant command.
- **Special Events.** The DOD provides support to a special event, a planned program of athletic competition or related activities involving participants from the US and/or other nations. Historic examples of these events are the Olympic Games and the Pan American Games.
- **Man-made Disasters.** Manmade disasters can be accidental or intentional. An example of an accidental event is an oil spill. The implications of a deliberate or unintentional large release of a CBRNE are severe. A catastrophic CBRNE event or attack may occur with little or no warning, resulting (either immediately or over time) in mass casualties and producing a mass exodus of evacuees. The response capabilities and resources of the local jurisdiction (to include mutual aid from surrounding jurisdictions and response support from the state) may be insufficient and quickly, if not immediately, overwhelmed.

Domestic Support Operations

4-131. Domestic disasters occur due to various causes and many may require the assistance of DOD and more specifically, the Army medical evacuation assets to fill response shortfalls. In the post-9/11 environment, terrorism or other intentional acts are recognized as a more likely cause of contingencies within the US than many had acknowledged since the civil defense era of the forties and fifties. This threat increases the likelihood that a large-scale disaster will occur that requires the augmentation of civil authorities' resources by DOD. The US DHS and USNORTHCOM have been established to coordinate the preparedness for and response to such events for civil authorities and DOD respectively.

4-132. Domestic support operations are typically conducted to prevent loss of life and property. Certain operations may be conducted in the form of immediate and automatic response by US military commanders as defined in DOD Directive 3025.15, to save lives, prevent human suffering, or to mitigate great property damage under imminently serious conditions. Other requests for support for civil disasters are made by civil authorities or a DCO (in a Stafford Act declared emergency) through the Joint Directorate of Military Support per the directive and work down through command channels from the Secretary of Defense.

4-133. Homeland Security Presidential Directive 5 recognized that there currently exists an artificial separation of local, state, and federal response to an incident and the role of DOD is not well streamlined or defined. This recognition led the White House and the DHS to initiate the transition from the Federal Response Plan to the National Response Plan, which serves as the doctrine for how the US responds to disasters no matter the cause.

4-134. Medical evacuation support plans to augment civil authorities are necessary for a variety of contingencies. Domestic support planning must address a range of problems such as—

- Preidentifying medical evacuation capabilities, units and personnel available to support various contingencies large enough to require civil support.
- Command and control relationships between civil authorities and DOD forces especially when DOD units are divided.
- Support for deployed DOD forces when no DOD logistics operations are deployed, including medical support.
- Cost capture and reimbursement from civil authorities to DOD in non-Stafford Act emergencies.

SECTION XI — OTHER TYPES OF MEDICAL EVACUATION SUPPORT MISSIONS

EVACUATION OF MILITARY WORKING DOGS

4-135. Military working dogs (MWDs) when injured or ill may be evacuated on any transportation means available. The using unit is responsible for the evacuation of the animal. Use of dedicated medical evacuation assets (air or ground ambulances) is authorized based on mission priority and availability. When possible, the handler should accompany the animal during the evacuation. Using units should include the location of veterinary treatment facility/support units on operational overlays. See FM 4-02.18 for information on veterinary support for MWDs.

PERSONNEL RECOVERY OPERATIONS

4-136. Air ambulances do participate in PR operations in a support role and are not protected from attack while engaged in a PR mission over contested or denied territory. If the involvement in these operations consists solely of evacuating wounded crewmembers from a crash site in friendly territory, AAs retain the protection accorded to them under the provisions of the Geneva Conventions. Air ambulances flying in contested or denied areas are not protected from attack unless flying at times and routes agreed to by the enemy and may be summoned to land. Air ambulances must obey a summons to land. Personnel retain their protections under the Conventions. However, if AAs participate in the actual search and rescue phases of the operation, they are not solely engaged in the provision of AHS and are, therefore, not afforded the protections. Further, because the mission does not fall within the protected activities, the AAs participating in the PR operation must remove the Geneva Conventions emblem from the aircraft.

SHORE-TO-SHIP EVACUATION OPERATIONS

4-137. It is imperative that Army AA units be able to interface on the first day of battle with US Navy (USN) air-capable ships. Lessons learned from past operations, have shown that US Army helicopters should be able to operate to and from USN air-capable ships. An interservice agreement between the Army and the Navy allows for deck-landing qualification of Army pilots (refer to FM 1-564 for additional information). It is important that units having contingency missions requiring Navy support establish training requirements to obtain naval-operations orientation, water egress training, water survival, and deck-landing qualification. This enhances the successful accomplishment of the aeromedical evacuation mission to naval vessels.

COMMUNICATIONS

4-138. In past joint operations, communications have been burdensome for both Army and Navy elements. It is essential to establish commonality of communication requirements during training exercises and to establish communication equipment and frequencies for medical evacuation to Navy vessels. This will provide smooth integration of Army helicopters into the Navy airspace management system during actual operations.

NAVIGATION

4-139. As Navy vessels may operate relatively long distances from the ground combat operations, Army AA units need to be proficient in over water navigation. The use of navigational aids (NAVAIDS) from the Navy element in support of the operation is the first priority for over water navigation. Basic dead-reckoning remains a secondary measure.

MEDICAL EVACUATION OF ENEMY PRISONERS OF WAR AND DETAINEES

4-140. Sick, injured, and wounded EPW and detainees are treated and evacuated in military police (MP) channels when possible. They must be physically segregated from US, allied, and coalition patients. Guards for these prisoners are provided according to the BCT, division or corps TSOP and are from other than medical resources. The echelon commander is normally responsible for this support. The US provides the same standard of medical care for wounded, sick, and injured EPW and detainees as that given to US, allied, and coalition Soldiers. Wounded, sick, or injured EPW and detainees in the CZ may be treated and returned to MP channels for evacuation; or the EPW and detainees may be stabilized and moved through medical channels to theater hospitals for treatment. Enemy prisoners of war and detainees are not evacuated from the theater for medical treatment.

4-141. When EPW are evacuated through medical channels, medical personnel—

- Report this action through medical channels to the next higher headquarters.
- Request disposition instructions from the MEDBDE IPMC.
- The MEDBDE IPMC is responsible for—
  - Coordinating the transportation means.
  - Identifying the MTF to which the EPW/detainees will be taken.
  - Coordinating, in conjunction with the hospital commander, with the Detainee Reporting System to account for EPW within medical channels.

SECTION XII — MEDICAL EVACUATION REQUEST

4-142. Procedures for requesting medical evacuation support must be institutionalized down to the unit level. Procedural guidance and standardization of request procedures are provided below. The same format used to request aeromedical evacuation is also used for requesting ground evacuation.

Table 4-3. Procedures for information collection and medical evacuation request preparation

Line	*Item*	*Explanation*	*Where/How Obtained*	*Who Normally Provides*	*Reason*
1	Location of pickup site	Encrypt the grid coordinates of the pickup site. When using the DRYAD Numeral Cipher, the same "SET" line will be used to encrypt the grid zone letters and the coordinates. To preclude misunderstanding, a statement is made that the grid zone letters are included in the message (unless unit SOP specifies it's use at all times)	From map	Unit leader(s)	Required so evacuation vehicle knows where to pickup patient. Also, so that the unit coordinating the evacuation mission can plan the route for the evacuation vehicle (if the evacuation vehicle must pick up from more than one location).
2	Radio frequency, call sign and suffix	Encrypt the frequency of the radio at the pickup site, not a relay frequency. The call sign (and suffix if used) of person to be contacted at the pickup site may be transmitted in the clear.	From SOI	RTO	Required so that evacuation vehicle can contact requesting unit while enroute (obtain additional information or change in situation or directions).
3	Number of patients by precedence	Report only applicable information and encrypt the brevity codes. A—URGENT B—URGENT-SURG C—PRIORITY D—ROUTINE E—CONVENIENCE If two or more categories must be reported in the same request, insert the word "BREAK" between each category.	From evaluation of patient(s)	Medic or senior person present	Required by unit controlling vehicles to assist in prioritizing missions.

4	Special equipment required	Encrypt the application brevity codes. A—None B—Hoist C—Extraction equipment D—Ventilator	From evaluation of patient/situation	Medic or senior person present	Required so that the equipment can be placed on board the evacuation vehicle prior to the start of the mission.
5	Number of patients by type	Report only applicable information and encrypt the brevity code. If requesting medical evacuation for both types, insert the word "BREAK" between the litter entry and ambulatory entry. L+# of patients–Litter A+# of patients–Ambulatory (sitting)	From evaluation of patient(s)	Medic or senior person present	Required so that the appropriate number of evacuation vehicles may be dispatched to the pickup site. They should be configured to carry the patients requiring evacuation.
6	Security of pickup site (wartime)	N—No enemy troops in area P—Possibly enemy troops in area (approach with caution). E—Enemy troops in area (approach with caution). X—Enemy troops in area (armed escort required).	From evaluation of situation	Unit leader	Required to assist the evacuation crew in assessing the situation and determining if assistance is required. More definitive guidance can be furnished to the evacuation vehicle while it is enroute (specific location of enemy to assist an aircraft in planning its approach).
6	Number and type of wound, injury or illness (peacetime)	Specific information regarding patient wounds by type (gunshot or shrapnel). Report serious bleeding, along with patient's blood type if known.	From evaluation of patient(s)	Medic or senior person present	Required to assist evacuation personnel in determining treatment and special equipment needed.
7	Method of marking pickup site	Encrypt the brevity codes. A—Panels B—Pyrotechnic signal C—Smoke signal D—None E—Other	Based on situation and availability of materials	Medic or senior person present	Required to assist the evacuation crew in identifying the specific location of the pick up. Note that the color of the panel or smoke should not be transmitted until the evacuation vehicle contacts the unit (just prior to its arrival). For security, the crew should identify the color and the unit verifies it.
8	Patient	The number of patients in each category need not	From evacuation	Medic or	Required to assist in

	nationality and status	be transmitted. Encrypt only the applicable brevity codes. A—US military B—US citizen C—Non-US military D—Non-US citizen E—enemy prisoner of war (EPW)	platform	senior person present	planning for destination facilities and need for guards. Unit requesting support should ensure that there is an English speaking representative at the pickup site.
8	CBRN contamination (wartime)	Include this line only when applicable. Encrypt the applicable brevity codes. C—Chemical B—Biological R—Radiological N—Nuclear	From situation	Medic or senior person present	Required to assist in planning for the mission. (determine which evacuation vehicle will accomplish the mission and when it will be accomplished)
9	Terrain description (peacetime)	Includes details of terrain features in and around proposed landing site. If possible, describe relationship of site to prominent terrain feature (lake, mountain, tower).	From area survey	Personnel present	Required to allow evacuation personnel to assess route/avenue of approach into area. Of particular importance if hoist operation is required.

Chapter 5

Medical Evacuation in Specific Environments

This chapter addresses medical evacuation in specific environments or under special circumstances. The medical evacuation effort must be well-planned and its execution synchronized to be effective. Further, medical evacuation personnel must be flexible and ready to improvise, if needed, to meet the demands of unique situations.

SECTION I — MOUNTAIN OPERATIONS

5-1. In the past, armies have experienced great difficulty in evacuating patients from mountainous areas because they are extremely diverse in nature. Some mountains are dry and barren with temperatures ranging from extreme heat in the summer to extreme cold in the winter. In tropical regions, mountains are frequently covered by lush jungles and heavy seasonal rains occur. Many areas display high rocky crags with glaciated peaks and year-round snow cover. Elevations can also vary from as little as 1,000 feet to over 16,000 feet with drastic and rapidly occurring weather changes.

5-2. Operations in mountainous terrain require some procedure modifications. This is due to the environmental impact on personnel and equipment. Important physical characteristics and considerations that influence medical evacuation are—

- Rugged peaks, steep ridges, and deep valleys.
- Limited number of trafficable roads.
- Reduced communications ranges.
- Unpredictability of and severe changes in weather.
- Decreased partial pressure of oxygen.
- Limited availability of LZs.

5-3. In order to effectively support the tactical plan, the HSS/FHP plan must provide maximum flexibility. The HSS planner should consider using all means of evacuation. Due to the length of evacuation times and the limited means of ground evacuation, it is important to triage and prioritize patients prior to movement.

- The availability of improved, hard-surfaced roads is extremely limited, if they exist at all. Usually, improved roads are only found in valley corridors. Such roads are often dependent upon a system of narrow bridges spanning mountain streams and ravines. They may also twist along ridgelines and cling to steep shoulders.
- Secondary roads and trails may be primitive and scarce. However, they may provide the only routes capable of vehicular traffic. Cross-compartment travel between adjacent valleys may be impossible by ground vehicle. Off-road travel requires detailed planning, even for short distances.
- Because of rough terrain, the Role 2 MTF may not be able to reach the BAS by ground vehicle. An ambulance shuttle system is established with an AXP for air and ground evacuation vehicles to meet litter bearers. Litter bearers and beasts of burden may be the only means of evacuation available. Any available personnel may be used as litter bearers (nonmedical personnel from supported units may be required to augment the litter bearer teams). Close coordination between Role 2 MTFs and BASs in establishing CCPs or AXPs is necessary to—
  - Reduce distance traveled by litter bearers.
  - Reduce evacuation time.

  - Conserve personnel.
  - Locate the best potential LZs for AAs.
- In mountainous areas, evacuation of patients by air is the preferred means. Air ambulances permit the rapid movement of patients over rugged terrain. For example, to travel a distance of only 6 kilometers on foot could take up to 2 hours, while flying time could be less than 2 minutes.
- Frequency-modulated radios are the principal means of communication in this environment. The ability to transmit is hampered by the limitations of line of sight transmissions.
- The briefing of ambulance drivers needs to be extensive, including detailed strip maps and overlays. Further, specific instructions on what to do in various situations should be covered (such as if the vehicle breaks down or the unit moves).

5-4. The mountain environment, with its severe and rapidly changing weather, impacts on aircraft performance capabilities; accelerates crew fatigue; and requires special flying techniques. Having to rely on continuous aviation support for a successful mountain operation is risky.

- Flying in mountainous areas requires special training. Both the terrain and the weather influence basic flying techniques and operational planning. Rugged, mountainous terrain complicates flight route selection. Direct routes can seldom be flown without exposing the aircraft to detection and destruction by the enemy.
- Important considerations for AA operations in mountainous areas are—
  - Density altitude which is the most important factor affecting aircraft performance. Density altitude combines temperature, humidity, and pressure altitude, and provides the basis for lift capability. Density altitude can vary significantly between the pickup point and the LZ because of the time of day and changes in elevation. Frequent performance planning updates are essential.
  - Unpredictable winds which can produce significant turbulence, wind sheers, updrafts, and downdrafts. This further increases the risk of a catastrophe in a seemingly routine mission. Adverse winds along with high density altitude demand current and accurate performance planning. Pilots must plan for greater margins of safety.
  - Ice can clog intake ports, thus starving the engine of air, or it can collect on rotor blades resulting in a significant loss of lift. Asymmetrical shedding can cause severe out-of-balance rotor conditions.
  - Visibility can be reduced by low clouds or fog greatly decreasing the ability to navigate or to avoid obstacles.
  - Lack of LZs is critical because the characteristics of mountain terrain do not usually afford adequate LZs. The terrain may only allow the aircraft to hover while loading patients on board.
  - Hoist operations can be expected in mountainous terrain requiring the use of the internal or external rescue hoist. Mounting the rescue hoist on the aircraft as standard equipment in mountain operations may be required. When possible, orientation and training sessions with supported troops should be conducted to help minimize the difficulty of such missions. Depending on the terrain, the forest penetrator may also be needed to accomplish the mission.
  - Enemy air defenses must be considered because when enemy air defense capabilities preclude using AAs in forward areas, they should be used to evacuate patients from AXPs or from Role 2 MTFs.
  - Ambulatory patients may be reported as litter patients in mountainous terrain because these patients may be unable to move unassisted over the rugged terrain. Once placed on the AA, their status may be upgraded.
  - Additional crew training for ground and air evacuation crews should include training and orientation in mountaineering skills, handling patients, and survival skills. Such training might include cold weather survival training (cold injury prevention, mountain [rock] climbing, and the use of ropes and vertical rescue techniques). Training may also include individual and unit

movement at high altitudes and techniques of patient evacuation by litter, emphasizing the use of pack animals (if available from the host country), and the improvised travois litter.

- Patient loading can be hazardous and care must be taken when loading patients where there is a great deal of slope to the LZ. Emphasis on approaching and loading the aircraft from the down-slope side of the aircraft must be reinforced.

DANGER

Approaching the aircraft from the up-slope side is hazardous.

5-5. Troops operating in mountainous areas are exposed to other injuries and illnesses that frequently occur in this environment. These conditions include—

- An increased rate of fracture, sprain, and dislocation injuries.
- Incidents of acute mountain sickness, high-altitude pulmonary edema, and cerebral edema caused by rapid ascent to heights over 7,500 feet.
- Cold weather injuries.
- Dehydration and heat exhaustion.
- Sunburns and snow blindness.
- Aggravated sickle cell anemia. Although this condition is not considered a mountain illness, personnel with the sickle cell trait can be seriously affected by the decrease in barometric pressure and lower oxygen levels found at higher altitudes.

5-6. The proportion of litter cases to ambulatory cases is increased in mountainous terrain, for even the slightly wounded may be unable to move unassisted over rough terrain. Litter relay stations may be required along the evacuation route to conserve the energy of litter bearers and to speed evacuation. It is important to be able to predict the number of patients that can be evacuated with available personnel. When the average terrain grade exceeds 20 degrees, the four-man litter team is no longer efficient and should be replaced by a six-man team. The average mountain litter team should be capable of climbing 120 to 150 vertical meters of average mountain terrain and return with a patient in approximately 1 hour.

5-7. Mountain operations may require medical personnel to carry additional equipment. Items such as ropes, pulleys, pitons, piton hammers, and snap links are all necessary for evacuating patients and establishing BASs. All unnecessary items of equipment including those for which substitutes or improvisations can be made should be left behind. Heavy tentage, bulky chests, extra splint sets, excess litters, and nice-to-have medical supplies should be stored. Such medical supplies, if stored, should be readily available for airdrop or other means of transport. Medical items that are subject to freezing should be safeguarded; they should not be exposed to the low temperatures experienced in mountainous areas.

5-8. For further information on mountain operations, refer to FM 3-97.6.

SECTION II — JUNGLE OPERATIONS

5-9. Health service support elements in a jungle environment retain the same basic capabilities as in other environments. Jungle operations, however, subject personnel and equipment to effects not found in other environments. The jungle environment degrades the ability to maneuver. Security problems are also increased and affect medical evacuation operations as much as they do the combat forces.

5-10. In jungle operations, combinations of air and ground evacuation units are used to maximize the patient evacuation potential. Using this dual system of evacuation ensures that the inherent limitations of one system can be compensated for by the other. Jungle variations affect the organizing, positioning, and securing of HSS/FHP. Due to the terrain, aerial resupply is usually a common practice. The responsiveness provided by aerial resupply requires fewer supplies to be stockpiled in the combat trains.

5-11. Jungle combat operations are characterized by ambushes and other unconventional warfare operations. The security threat caused by infiltrators requires that LOC be patrolled often and that convoys be escorted. It is, therefore, essential that HSS be performed as far forward as the tactical situation permits. Positioning assets forward—

- Improves response time.
- Reduces road movement.
- Allows the HSS elements to take advantage of the security offered by combat units.

5-12. The thick foliage often makes evacuation by ground more difficult than in other types of terrain. Factors such as the threat, limited road network, and reliance on nonmedical personnel for convoy security make air evacuation the preferred means. By using the ambulance shuttle system, patients can be transferred from forward operating ground ambulances to either ground or AAs operating further to the rear. In situations where evacuation assets are delayed by various factors (weather or terrain), patients are held for longer periods of time at forward locations. This will dictate the need for additional medical supplies. Health service support planners must try to anticipate these delays whenever possible. The increased disease and infection incidences associated with the jungle environment may worsen the patient's condition; therefore, timely evacuation is essential.

5-13. In some remote and densely foliaged jungles, the only means of evacuation may be by litter. Ambulances may not be practical on trails, unimproved muddy roads, or in swamps. As in mountain operations, there is a higher proportion of litter cases than usual. In the jungle even a slightly wounded Soldier may find it impossible to walk through dense undergrowth. At best, litter teams can carry patients only a few hundred meters over rough jungle terrain before needing rest or relief. Litter carries should be kept as short as possible and medical elements prepositioned and retained forward.

5-14. Other special planning considerations in jungle operations include—

- Water is vital in the jungle and is plentiful. Water from natural sources, however, should be considered contaminated. Water purification procedures must be taught to all Soldiers. (Refer to FMs 21-10 and 4-25.12 for additional information.)
- Due to the tropical climate, units should pack hot weather clothing when deploying to jungle areas. Insect (mosquito) nets, insect repellent, permethrin treated uniforms and sunscreen should be issued to all Soldiers operating in this environment.
- The jungle environment is ideal for the transmission of a large number of diseases. The rate of DNBI casualties is potentially the highest in this climate. The heat, humidity, and terrain places the troops at high risk for dehydration, heat injury, skin diseases, endemic diseases, and immersion foot. Small wounds can rapidly become infected and lead to loss of effectiveness and possibly require evacuation. High standards of personal hygiene must be taught, encouraged, and maintained by the command. Mosquitoes and other arthropods that carry disease flourish under jungle conditions. Use of all possible personal protective measures must be ensured. Food- and waterborne diseases leading to diarrhea or other symptoms will abound. Food service sanitation measures must be strictly followed. The potential for contamination of food and water increases with each time they are handled, stored, or transported. Soldiers must be encouraged to consume adequate amounts of water that has been purified and to eat only approved foods. In the jungle it is necessary for the commander to pay meticulous attention to the details of PVNTMED measures to maintain an effective fighting force. For additional information on PVNTMED measures, refer to FM 21-10 and FM 4-25.12.
- Health service support personnel should be trained in survival and support techniques in jungle environments. For example, training should be conducted in—
  - Hot weather acclimatization and survival.
  - Prevention, early detection, and treatment of arthropod-, food-, and waterborne diseases.
  - Land navigation in a jungle environment.
  - Field sanitation and other PVNTMED measures.
  - Care and maintenance of equipment and supplies.

- Due to the increased heat and humidity, vehicles and equipment require additional maintenance. Equipment tends to rust quickly and must be cleaned and oiled more frequently. Canvas items rot and rubber deteriorates much faster than in more temperate climates.
- The range of FM communications in the jungle is significantly reduced due to the dense undergrowth, heavy rains, and hilly terrain. The range of a radio set operated in the jungle may be reduced by 10 to 25 percent. The heavy rain and high humidity of the tropics also reduce the range (about 20 percent) and reliability of wire communications. The transmission range can be extended by using additional radio relays and field expedient antennas.
- Utility helicopters are not able to lift the same size loads that can be lifted in more temperate areas. This results in a reduced patient load in some evacuation aircraft. Again, frequent and accurate performance planning is essential for mission accomplishment.
- There may be few suitable LZs. Many LZs will only be large enough to support one or two helicopters at a time.
- Hoist operations may be required more frequently in the thick jungle vegetation where LZs are not available. The forest penetrator should be carried on all operations.
- For aviation-specific information, refer to training circular (TC) 1-400.

SECTION III — DESERT OPERATIONS

5-15. Deserts are arid, barren regions of the earth incapable of supporting normal life due to a lack of fresh water. Although deserts are often thought of as hot climates, it is important to note that temperatures range from over 136 degrees (°) Fahrenheit (F) in some deserts, to bitter cold in others. Day to night fluctuations in temperature can exceed 70° F. Desert terrain can have mountains, rocky plateaus, or sandy dunes; some desert areas may contain all of these characteristics. Rain, when it falls, often causes flooding in low-lying areas. Winds can have a devastating effect upon AHS operations by destroying equipment and supplies and causing dust storms. Dust storms make navigation and patient treatment difficult. Since deserts vary considerably in the type of terrain and temperature, and in their cultural makeup, current medical intelligence should be obtained prior to deployment on operations conducted on desert terrain.

5-16. People have lived and fought in desert areas for thousands of years. However, the environmental effects on personnel can be extreme, especially for Soldiers not prepared for these operations.

- To be effective, Soldiers must be properly acclimatized to the desert. Two weeks are usually required to satisfactorily acclimatize troops to hot environments, using progressive degrees of heat exposure and physical exertion. Other potential acclimatization problems that may be encountered are the effects of dry air and altitude on the respiratory system. Since many desert areas are located in mountainous terrain, Soldiers must be acclimatized for both the altitude and the temperature. In some areas of the world, such as the Gobi Desert in East Asia, people must be acclimatized to the cold in addition to the dryness. (For additional information, refer to FM 21-10, FM 8-250, and TB MED 507.)
- Units deployed in desert areas typically have long LOC and are widely dispersed. As unit elements become more dispersed, commanders must rely more heavily on junior officers and NCOs to ensure that C2 is maintained and that discipline and PVNTMED programs are enforced. For a unit to be effective, a high level of discipline must exist at all levels of the organization.
- Water is the most basic need in a desert. Without it, Soldiers cannot function effectively for more than a few hours. Thirst is not an adequate indicator of the need for water. It is necessary for each commander to establish and enforce a supervised drinking program. Experience has shown many times that Soldiers do not drink enough fluids unless forced to do so. It is important to cool the water, if at all possible, to make it more appealing. Water supplies should be carefully guarded against accidental loss, sabotage, or contamination. Extra water must be carried by AHS vehicles for patients to drink and to cool heat casualties.
- Soldiers deployed in the desert are susceptible to endemic diseases and environmental injuries. Proper water discipline, vaccines, prophylactic measures, field sanitation measures, personal

hygiene, and other PVNTMED measures can reduce these risks. Cold weather injuries, heat injuries, and respiratory disease can also be prevalent. Proper clothing, equipment, and a water discipline program must have command emphasis in desert operations.

- Winds may very easily damage materiel such as aircraft, antennas, and tents. Equipment is protected by using covers, tie-downs, and shelters. Terrain helps shield equipment from the wind if site selection is done carefully. In some cases, special tools, such as extra long metal tent stakes, are necessary.
- The effects of wind and sand are interrelated. Desert sand starts to become airborne when the wind reaches about 20 knots. Sandstorms—
  - Restrict visibility.
  - Pose a hazard to eyes (especially for Soldiers wearing contact lenses).
  - Can contaminate water supplies (if they are not protected).
  - Make navigation difficult.
- The sun may cause sunburn of the skin and eyes (if protection is not used).

5-17. Seven characteristics of the desert environment that may adversely effect equipment are—

- Terrain. Trafficability varies with the type of terrain covered. Open, flat, and rocky terrain affords higher trafficability than do mountainous areas, lava beds, or salt marshes. Drivers must be well-trained in judging the terrain over which they are driving to select the best alternative routes.
  - Tracked vehicles are best suited for desert operations. However, they can throw tracks when traversing a rocky area. Their use is also limited in rough terrain with steep slopes.
  - Wheeled vehicles may be used in desert operations; however, they normally have a lower average speed than tracked vehicles and a higher incidence of damage and malfunction. Wheeled vehicles often bog down in sandy areas and cannot traverse many of the rougher areas.
  - In planning for desert operations, vehicles should carry extra repair parts (fan belts, tires, and other items apt to malfunction).
- Excessive heat causes vehicles to overheat, leading to greater than normal wear. The frequency of leaks on vehicles and aircraft is greater than in some other environments. Engine and transmission seals tend to dry out and crack; fuel lines wear out quickly; and water requirements for cooling vehicle engines are greater. Loss of water, through evaporation, must be included in logistical planning. Using vehicle and aircraft covers reduce the effects of heat while vehicles and aircraft are not in use. Refer to FM 1-202.
  - Batteries do not hold their charge efficiently in intense heat. Dry battery supplies should be increased to compensate for a higher usage rate.
  - Communications equipment must be protected from the heat in the desert. Dust covers are used on this type of equipment. If the equipment has ventilating ports, these should be cleaned regularly to avoid clogging.
  - Medical supplies must be protected from the heat to prevent deterioration. The shelf life of some medical supplies decreases when stored in hot climates.
  - Medical equipment must be protected from the effects of heat. It may be protected using the same techniques as those used to protect communications equipment.
- The sun burns unprotected skin and it may damage unprotected eyes. Soldiers should dress in loosely fitting clothing, use sunburn cream or oils to protect exposed skin, and wear sunglasses or goggles to protect their eyes. Soldiers should remain fully clothed. Removing clothing increases direct exposure of the skin to the sun and eliminates the beneficial cooling effects of the moisture trapped in clothing. Radiant light or its heat effects may be detrimental to plastics, lubricants, pressurized gasses, rubber, and some fluids. All vehicles and aircraft should be kept well-ventilated, and windshields should be covered to reduce heat buildup inside. Supplies of all types should be stored in a well-ventilated, shady area. Placing supplies in covered holes in the ground may reduce adverse heat effects.

- Dust and sand present one of the greatest dangers to the proper functioning of equipment. Sand mixed with lubricants forms an abrasive paste. Lubrication fittings, bearings, and filters should be inspected frequently and changed when required.
  - Aircraft should not be exposed to dust and sand any more than is absolutely necessary. Ground handling instead of hovering reduces sand ingestion. Dust and sand increase failure of microphone switches, signal distribution panels, and circuit breakers, and cause electrical motors and generators to burn out. Wheel and flight control bearings require more frequent cleaning; engines should be flushed frequently.
  - Medical and communications equipment may be adversely affected by dust and sand. Over a period of time, electrical insulation is damaged by windblown sand. When combined with the effects of lubricants on the insulation, dust and sand can become a major communications problem. Special care should be taken to brush dust off radio equipment and to keep ventilating ports and channels clear.
  - Sand can accumulate in airframes, on the bottom of armored vehicles, and in bearings on all types of equipment. This accumulation, combined with oil and condensation, adds extra weight to aircraft and vehicles as well as jamming their control linkages. Sand and grease buildups must be removed from bearings to ensure safe operation and control of aircraft and vehicles.
  - Dust trails created by hovering aircraft or ground vehicles can be seen in excess of 10 miles on a relatively flat desert. This exposes these assets to direct and indirect enemy fires. Ground vehicles should reduce their speed to the point that they do not create a dust signature.
- Humidity is a factor in some desert areas of the world, especially in the Middle East. Humidity can become a problem for short periods of time in other desert areas. Light coats of lubrication can help prevent rust; however, these benefits should be weighed against the dust-gathering qualities of oil. Demisting equipment is used on optics and night vision equipment to combat the effects of humidity.
- Temperature variation can cause condensation in humid desert areas affecting optics, fuel lines, air tanks, and weapons. Expansion and contraction of air and fluids cause tires to over inflate during the day and under inflate at night. Fuel tanks may overflow during the day causing a fire hazard. Oil fluid levels become overfull and cause leaks during the day, or insufficient lubrication occurs when the oil cools. Vehicle operators and crew chiefs must ensure that the effects of temperature variations do not become a significant problem.
- Static electricity is prevalent in the desert. This is important to remember during refueling operations and when oxygen is being used onboard vehicles or aircraft. Proper refueling procedures must be followed. Static electricity also causes severe shock to ground personnel in sling load and hoist operations.

CAUTION

During hoist and sling load operations, the load must touch the ground before the ground crew can handle it.

5-18. To ensure success in desert operations, detailed planning is required. Factors to consider include—

- Additional quantities of water are required for HSS/FHP operations for the survival of both medical personnel and their patients. Load plans for all vehicles and aircraft must include water. Water is as mission essential as any piece of unit equipment. It should be a priority item when load plans are developed.
- Prescribed load lists are expanded to carry sufficient quantities of repair parts easily degraded by the environmental factors. For example, rubber and plastic fittings and tubes, or spare parts for communications equipment.

- Plan for the effects of wind, sand, and sun on all equipment. All plastic and glass surfaces on vehicles, aircraft, and other equipment should be covered when not in use. Covers should be ordered or made prior to deployment.
- Fuel planning is critical due to power limitations, extended range requirements, and increased vulnerability of refueling sites in the relatively open desert terrain. Careful planning of FARPs is essential for mission accomplishment.
- Units should plan to pack both hot and cold weather clothing when deploying due to varying temperatures between day and night.
- Petroleum, oils, and lubricants (POL) should be of the proper viscosity for desert operations. Maintenance services are also performed more frequently on ground vehicles and aircraft, thus requiring a larger amount of POL than normal.
- Extra filters of all types are planned for due to a higher consumption rate.

5-19. Training for desert operations is not significantly different than training for operations in other areas except for the following:

- Because many desert areas are in mountainous terrain and because high temperatures increase density altitude, aeromedical evacuation units should conduct mountain training to prepare for contingencies in desert areas. Further, procedures and techniques for evacuation in mountainous terrain must be practiced by all AHS personnel. Special equipment requirements (paragraph 5-7) must also be considered.
- Navigation in desert terrain varies from simple to extremely difficult. Factors affecting navigation are the type of desert and the scale and quality of the available navigational charts. At times, aircraft may have to use dead-reckoning navigational techniques (time, distance, and heading). Ground vehicles must have compasses available, as they have to rely on compass headings and odometer readings to navigate. Ground and AA crews should be able to interpret navigational charts and maps of all types and scales. Use of convoys is a viable technique to improve security and to ensure that ground vehicles do not get lost. Aircraft may be used to assist in navigation by convoys in those areas in which there are poor road networks and the terrain offers no distinctive features by which to navigate.

5-20. In principle, medical evacuation operations in the desert do not differ greatly from these operations in other environments.

5-21. Helicopter landing sites should be chosen with care. Common mistakes made by many units when establishing the LZ are—

- Locating the pad relative to the patient and tents, vehicles, and other obstacles. A common tendency is to locate the helipad downwind of MTFs so that approaches may be made into the wind towards the facility. In mountainous deserts, winds normally channel down the valleys and are more predictable along valley floors. A better site selection for an LZ is with the MTF, alongside the approach and takeoff zone. Thus, the landing direction is up or down the valley, depending on the airflow, and the MTF is not over flown.
- Situating landing sites in washes, small confined areas between large rocks, or close to moving tracked vehicles. Map coordinates are rarely accurate unless the site is beside a major terrain feature. Therefore, LZs should be located next to major terrain features or on higher ground where they can be seen from the air at a distance of 2 to 3 kilometers, if possible. Lack of distinctive features in the open desert and on large-scale maps makes pinpoint navigation difficult, especially at night.
- Marking of helicopter LZs is done so that the pad can be seen from the air, but the markings should not be a hazard in themselves. If engineer tape is used, it should be firmly secured to prevent it from blowing loose. Panel markers are not a good tool to use as they are difficult to see. If panel markers are used, they need to be secured. If used, flares or marker smoke should not be deployed on or directly upwind from the pad. Smoke grenades or flares should not be thrown under the aircraft as it lands. Avoid using white smoke to mark the LZ. Colored smoke is probably the best daylight marking method. It is difficult to detect a smoke grenade more

than 2 to 3 kilometers away, but an aircraft in the general vicinity can normally see it. Radios are used to guide aircraft to the LZs, but this creates an electronic signature. Units requesting medical evacuation must be prepared to signal the evacuation aircraft upon its arrival. Normally, map coordinates will guide the aircraft to within 2 to 3 kilometers of the LZ. Even from nap-of-the-earth (NOE) altitudes, the aircrew may be able to see several units in the area. The requesting unit must signal the aircraft to ensure the designated LZ is used.

5-22. When planning for night flights, considerations include the following—

- Moonlight aids the medical evacuation pilot by providing him with the light to see with either unaided vision or night vision goggles (NVGs). When adequate ambient light exists, medical evacuation crews function almost as effectively at night as they do during daylight. The small arms threat is somewhat reduced at night, although it still exists from radar-guided weapons, infrared-sited weapons, and passive night vision device-equipped weapons systems. Flying into a bright moon with NVGs on can be compared to flying into the sun during the day. The goggles darken and visibility becomes extremely poor. Flight routes should not be planned to fly directly into a bright moon if NVGs are to be used.
- The lack of visual cues over sand is similar to that over water. It is very easy for pilots to become disoriented and fly into the ground. Reliance on radar altimeters is a must over flat sandy areas of the desert.
- Frequently, desert areas do not have sufficient ambient light to allow adequate night vision, even with the aid of NVGs. A pilot wearing NVGs is often unable to see the ground at an altitude of 100 feet using a landing light equipped with a pink light filter. Under these conditions, dead reckoning is the only effective navigation method unless global positioning satellites, Doppler equipment, or NAVAIDS are available. Unfiltered light can be used with or without NVGs; however, this increases the risk of exposing the aircraft's position to the enemy.
- Desert warfare is usually characterized by extended battle zones which increase evacuation distance and time. Combat health support units are located further to the sustainment area in the desert. Establishing an ambulance shuttle system or CCPs is useful. Combat health support units require a greater number of vehicles for operating in deserts than in other environments. Air evacuation by fixed and rotary-wing aircraft is the preferred method due to their speed and range. Further, using aircraft reduces the load on ground vehicles. Augmentation from higher echelon HSS may also be required to meet the extended evacuation needs.
- Smoke is used extensively on the modern battlefield by both sides. It can be effectively used to mask friendly actions to include medical evacuation. (Refer to Appendix E for further information.)
  - Smoke can be a major hazard, especially to medical evacuation helicopters. Smoke reduces visibility and forces an aircraft higher where it can be acquired by threat weapons systems. The phenomenon of inversion occurs often in the desert. When this happens, medical evacuation vehicles and aircraft may be able to work underneath the smoke using the smoke layer for overhead concealment.
  - Medical units must coordinate closely with supported organizations on smoke operations. Smoke can either help or hinder the evacuation mission, depending upon how it is used.
- Communications in the desert are affected by a number of factors. Atmospheric interference and the skip of signals occur frequently. Mineral deposits in the desert may unexpectedly disrupt communications. Many of these problems can be overcome by using additional radio relays, pre-established control measures, and visual signals.
- Artificial lights may be used at times in the desert. They are very easily detected. Even with blackout lights, vehicles using lights can be detected for miles with NVG. Serious consideration should be given to driving without using lights when the tactical situation dictates. Ground guides are used to help vehicles navigate through areas that are not clearly marked or through area where troops are present.
- Wind is one of the most significant environmental factors affecting medical evacuation in the desert. Wind can be destructive to both structures and equipment; tents, antennas, and aircraft

can be easily damaged. Wind direction and speed vary greatly within the space of a few miles. Velocity is substantially increased when wind channels between hills and direction changes due to interference of terrain features. The wind frequently makes aeromedical evacuation impossible by exceeding the operating limitations of the aircraft. At other times, it may limit the use of some potential LZs. Blowing sand and dust can slow down the evacuation system by making navigation by either ground or AA difficult, if not impossible. High winds are predictable to a certain extent. For example, at certain times of the year in the Mojave Desert high winds occur every day at dusk and last for 3 to 4 hours. At other times, high winds, based on frontal weather patterns, can remain for several days at a time. These factors should be considered by HSS planners, and medical evacuation assets should be massed or relocated accordingly.

- The desert provides little or no protection from enemy air defenses except in mountainous terrain. Aircraft may have to be flown above NOE altitudes to prevent a dust signature. These factors cause increased exposure and vulnerability of AAs to enemy air defenses and may limit their employment.

5-23. Further Information. Refer to FM 90-3 for additional information on desert operations.

SECTION IV — EXTREME COLD WEATHER OPERATIONS

5-24. Operations in the extreme cold have many of the limiting factors found in desert operations. The tundra and glacial areas are harsh, arid, and barren. Temperatures may reach lows of -80° F to -100° F which, combined with gale force winds, make exposure unsurvivable.

- The greatest environmental detriment to operations is blowing snow. This results in a loss of depth perception from total white conditions. Blowing snow is caused by the wind or by the rotor wash of helicopters; its effect reduces visibility to zero.
- Other environmental considerations are as extreme but easier to circumvent. Solid footing is suspect in both the dead of winter and in the summer. Snow and ice cover crevasses, holes, and otherwise unstable ground. In traversing suspect ground situations, consider linking Soldiers by rope. During the summer, ground transportation is more restricted than in any other environment due to the marsh and muskeg composition of the arctic tundra. Patients must be sustained for a longer duration due to terrain delays and the lack of direct lines of evacuation.
- Greater responsibility has to be placed on each Soldier, especially for maintenance of nutrition and water consumption. It is imperative to stress that leadership and training are important in the prevention of cold weather injury. Strict adherence to the guidelines found in FMs 21-10 and 31-70 assures an effective fighting force. Water conservation is essential; however, adequate consumption by the individual should be enforced.
- Factors to consider for conducting evacuation in arctic operations include the following:
  - Arctic warfare is usually characterized by extended battle zones that increase evacuation distance and time. Establishing an ambulance shuttle system or CCPs is useful. Augmentation from higher echelon AHS may also be required to meet the extended evacuation needs.
  - Additional supplies of water should be carried by ambulances and maintained at CCPs, if possible.
  - Due to the decreased temperature and frozen environment, ambulance maintenance requirements are increased. Lubricants must be of the correct viscosity for the temperature. In extreme cold, batteries perform less efficiently. Consult the specific aircraft maintenance manual for the recommended battery and procedures for extreme cold weather operations.

Note. All ambulances are considered deadlined without a functional heater for the patient compartment.

  - The proper storage of medical supplies is essential to prevent loss from freezing.

- There are few terrain features or road networks; therefore, evacuation routes must be surveyed and marked over open terrain. At extreme latitudes, operations during the winter months are conducted in extended hours of darkness. The use of NVG may be required. Compass accuracy is inconsistent due to a geomagnetic phenomenon. Beacons and homing devices are essential for air navigation.
- Weather is extremely unpredictable. There are too few observers to allow for accurate assessment of weather patterns. Unfavorable weather conditions cause unexpected delays; therefore, medical personnel must be prepared to provide survival measures for their patients and themselves.
- Landing zones must be chosen with extreme care in both winter and summer. Blowing snow mandates instrument-assisted takeoffs and running landings. Landing areas must be correspondingly larger. The full weight of the aircraft cannot be allowed to settle on the skis until after firm ground conditions are established. Movement of patients to and from the aircraft is difficult. Where an aircraft lands is where it stays. A rocking motion, to free the skis prior to lift off, is performed using the cyclic and antitorque controls.

- Thorough planning and strict preparation are the keys to survival. Factors to consider include the following:
  - Mud obstacles at noon may become an avenue of approach at midnight.
  - Snow complicates all work. Snow-covered terrain hampers reinforcements, muffles noise, makes cross-country driving hazardous, and creates different camouflage requirements.
  - Because of thermal sights, a complete reappraisal of concealment is required.
  - Tracks in the snow destroy concealment.
  - No Soldier is assigned to any job alone. The buddy system is used at all times.
  - Anticipate that all maintenance tasks will take twice as long.
  - Bare metal can stick to skin or wet garments in subfreezing temperatures.
  - Fuel spilled on skin or garments increases the freezing factor; it is one of the greatest causes of injury in winter operations.
  - When operating in the cold, anticipate increased POL needs. Fuel consumption can rise as much as 25 percent for vehicles operating in deep snow, slush, or mud.
  - The recommended fuel for Yukon stoves is diesel.
  - Make every effort to warm gearboxes and engines before starting.
  - The first consideration in the AO is heat; followed by shelter for sustained work.
  - Soldiers need to stand clear of taut cables; steel tends to be brittle and breaks in extremely cold temperature.
  - Fire extinguishers are winterized by adding 15 percent nitrogen to the carbon dioxide.
  - Degradation of battery life requires changes as much as six times more frequently than in a more temperate environment.
  - Radio sets are warmed up prior to transmission. The sets may be turned on but should not transmit for at least one-half hour.

Note. Single-channel ground and airborne radio system (SINCGARS) radios do not require a warm up period.

  - Frost shields (such as using the plastic bag in which the batteries are packed) should be placed over microphones.
  - Grounding rods have to be buried horizontally instead of pounded in vertically. Recovery of stakes and rods placed in the ground is significantly more difficult.
  - Flooring is needed in heated areas because of the thawing of the tundra.
  - Soldiers must take breaks for water and warmth.
  - Static electricity presents a serious safety hazard especially around flammable materials.

5-25. For additional information, refer to FM 31-70.

SECTION V — CHEMICAL, BIOLOGICAL, RADIOLOGICAL, AND NUCLEAR ENVIRONMENTS

5-26. Medical evacuation and treatment are conducted continuously throughout operations conducted in a CBRN contaminated environment. The AHS commander must have a comprehensive plan which is rehearsed on a periodic basis to ensure the timely evacuation and treatment of casualties in a CBRN environment. Techniques and procedures which are essential for operating in a contaminated environment should be contained in the unit TSOP. The number of casualties and their medical condition, type of contaminant, the size of the land area contaminated, the expected duration of operation, risk assessment and acceptable level of risk, and the number of AHS assets (medical personnel, medical units, and evacuation vehicles and aircraft) initially contaminated will determine the quantity and type of uncontaminated AHS resources, if any, which will be introduced into the contaminated environment to ensure timely medical treatment and evacuation occur.

5-27. The Commander must take into consideration the number of assets he is willing to commit during evacuation operations in a CBRN contaminated environment. Since the combinations of evacuation methods are nearly endless, the commander has greater flexibility in tailoring an evacuation plan to meet his particular tactical situation in a CBRN contaminated environment.

5-28. On the modern battlefield there are three basic modes of evacuating patients (personnel, ground vehicles, and aircraft).

- In using personnel to physically carry the casualties, the commander must realize the inherent stress involved. Cumbersome mission-oriented protective posture (MOPP) gear needed in a contaminated environment (added to climate, increased workloads, and the fatigue of battle) greatly reduces the effectiveness of unit personnel.
- If the commander must send evacuation personnel into a radiological contaminated area, he must establish operational exposure guidance (OEG) for the medical evacuation operation. Radiation exposure records are maintained by the unit CBRN NCO and are made available to the commander, staff, and surgeon. Based on OEG, the commander decides which medical evacuation assets to send into the contaminated environment.

5-29. Commanders should make every effort to limit the number of evacuation assets which are contaminated while still maintaining a timely and effective medical treatment and evacuation operation.

- It is expected that a certain number of both ground and AAs will become contaminated in the course of battle. The commander can, therefore, segregate the contaminated ones. This results in the smallest impact on his available assets and the greatest possibility for continuing the patient evacuation mission. Optimize the use of resources, medical or nonmedical, which are already contaminated before employing uncontaminated resources.
- Once a vehicle or aircraft has entered a contaminated area, it is highly unlikely that it will be able to be spared long enough to undergo a complete decontamination. This depends upon the contaminant, the tempo of the battle, and the resources available. Normally, contaminated vehicles (air and ground) have restricted use and are confined to dirty environments.
- Introducing uncontaminated aircraft into a contaminated area should be avoided, whenever possible. Ground ambulances should be used instead of AAs as long as their use does not adversely affect the patient's medical condition. Ground ambulances are more plentiful and are easier to decontaminate. This does not, however, preclude using aircraft in a contaminated environment or in the evacuation of contaminated patients.
- The relative positions of the contaminated area, FLOT, and threat air defense systems determine if and where helicopters are to be used. The commander may choose to restrict one or more helicopters to the contaminated areas and use ground vehicles to cross the line separating contaminated and clean areas. The ground ambulance can proceed to the receiving MTF with a

patient decontamination station. The patient can then be transferred to a clean ground or AA if further evacuation is required. The routes used by ground vehicles to cross between contaminated and clean areas are considered dirty routes and should not be crossed by clean vehicles. The effects of wind and time upon the contaminants must also be considered.

- The rotor wash of the helicopters must always be kept in mind when evacuating contaminated casualties. The intense winds disturb the contaminants in the area and further aggravate the condition by additionally spreading the contaminants. Ideally, the aircraft must be allowed to land and reduce to a flat pitch prior to bringing any patients near. This will be dictated by the tactical situation, but allows some reduction in the effects of the downwash. A helicopter must not land too close to a decontamination station (especially upwind) because any trace of contaminants in the rotor wash will compromise the decontamination procedure.
- Immediate decontamination of aircraft and ground vehicles should be accomplished to minimize crew exposure. Units should develop their own procedures for thorough decontamination and document them in their TSOPs. A sample aircraft decontamination station that may be tailored to a particular unit's needs is provided in FM 3-11.5.
- Evacuation of patients must continue even in a contaminated environment. The commander must recognize the constraints placed upon him by resources and plan and train to overcome deficiencies.
- Refer to FM 4-02.7 for additional information on AHS operations in a CBRN environment.

SECTION VI — SHORE-TO-SHIP EVACUATION OPERATIONS

5-30. It is imperative that Army aeromedical evacuation units be able to interface with USN air-capable ships. Lessons learned from past operations, such as Vietnam and Grenada, have shown that US Army helicopters should be able to operate to and from USN air-capable ships. An interservice agreement between the Army and the Navy allows for deck-landing qualification of Army pilots. (Refer to FM 1-564 for additional information.)

- It is important that units having contingency missions requiring Navy support establish training requirements to obtain naval-operations orientation, water egress training, water survival, and deck-landing qualification. This enhances the successful accomplishment of the aeromedical evacuation mission to naval vessels.
- In past joint operations, communications have been burdensome for both Army and Navy elements. Commonality of communication requirements should be established during training exercises. Communication equipment and frequencies for medical evacuation to Navy vessels must be established. This will provide smooth integration of Army helicopters into the Navy airspace management system during actual operations.
- As the Navy vessels may operate relatively long distances from the ground combat operations, Army aeromedical evacuation units need to be proficient in over water navigation. The use of NAVAIDS from the Navy element in support of the operation is the first priority for over water navigation. Basic dead-reckoning remains a secondary measure.

5-31. An important aspect of joint operations is the medical capabilities of Navy vessels servicing the CZ. Knowledge of ship's medical capabilities assists the MRO to direct patients to proper treatment sites. There are many classes of ships which can meet the medical needs of ground forces. Destroyer tenders and aircraft carriers have helicopter landing areas, one operating room (OR) and, at a minimum, one medical officer. Amphibious ships have the most extensive medical facilities of any Navy combat ship. The Navy has 59 amphibious ships in active commission plus two tank landing ships, which are operated by the Navy Reserve Force. The primary mission of the amphibious ships is to transport and support the Fleet Marine Force. The ships have the additional duty of casualty receiving and treatment ships (CRTS). During normal operations, the medical staff is kept to a minimum. The medical staff is augmented when expanded capabilities are needed. Current information regarding landing requirements and medical capabilities should be obtained during training periods with the Navy. Casualty care is secondary to the combat mission of all US combat ships.

- Amphibious assault ships "WASP" class are designated by the Navy as amphibious helicopter assault carrier dock (LHD) (followed by a number) and have the largest patient care facilities on any US combat ship. The WASP class ships have six main ORs, four dental ORs, bed capacity that can be expanded to 600, and it carries 1,500 pints of frozen blood. This ship can receive casualties from helicopters or landing craft.
- Amphibious assault ships "TARAWA" class are designated by the Navy as amphibious helicopter assault carrier (LHA) (followed by a number). The TARAWA class ships have three main ORs, two dental ORs, an overflow bed capacity of 300, and carries 1,500 pints of frozen blood. The ship can receive casualties from helicopters or landing craft.
- Amphibious assault ships "IWO JIMA" class are designated by the Navy as amphibious assault helicopter carrier—aviation (LPH) (followed by a number). These ships were specifically designed to operate helicopters. The IWO JIMA class ships have two ORs and an overflow bed capacity of 200.
- The amphibious transport dock (LPD) (followed by a number) has less medical capabilities than the LHD, LHA, or LPH ships. It can be designated as secondary casualty receiving ship.
- The older dock landing ship (LSD) (followed by a number) can be used as a secondary CRTS when augmented. The newer class of LSD currently under construction can be used as a casualty receiving ship with a capacity for 50 wounded.
- The tank landing ship (LST) (followed by a number) is another type of ship used in amphibious operations. It is designed with a helicopter platform and a stern ramp. Patients can be delivered by air or boat when required by tactical or mass casualty situations. When the LST is augmented with medical personnel and materiel, it can be used for the emergency treatment and evacuation of patients.
- The troop transport (AP) (followed by a number) is not in active service. When available, the troop transport can be outfitted with special medical facilities and carry sick, injured, and wounded personnel.

5-32. The Military Sealift Command operates two hospital ships. The United States Navy Ship (USNS) MERCY T-AH 19 and the USNS COMFORT T-AH 20. One ship is based on each coast and, when needed, will be assigned medical staffs from military hospitals, getting underway within 5 days. The hospital ships MTFs were designed for a total capacity of 1,000 casualties, including 500 acute care beds and 500 recuperation beds. The hospital ships have 50 trauma stations in the casualty receiving area; 12 operating rooms; a 20-bed recovery room; 80 intensive care beds; and 16 intermediate, light, limited care wards. The maximum patient flow rate, for which the helicopter facility and the casualty reception area were designed, is 300 patients per 24 hours. There is a limited capability to receive casualties from boats.

5-33. The US Army has the shore-to-ship medical evacuation mission on an area support basis for Marine forces deployed on land.

SECTION VII — AIRBORNE AND AIR ASSAULT OPERATIONS

5-34. Airborne and air assault operations are some of the most complicated military operations to undertake, not only from an operational perspective, but also from a sustainment and conservation of the fighting strength perspective. A number of personnel within the medical chain that need to coordinate, not only amongst the AMEDD, but also include the aviation community as what proponent own the aircraft (for example, BAE, BSS, medical platoon, and such). Manual evacuation techniques may have to be used and planners should plan for redundant communications, only ground medical evacuation may or may not be available.

5-35. The airborne and air assault operational forces are specialized forces employed to maximize their design characteristics. Airborne units are a flexible force that can be strategically or tactically deployed. They can be inserted rapidly anywhere in the world as either a deterrent or strike force. Air assault units are flexible and lethal fighting organizations. They are ideally suited for rapid employment to critical areas beyond the reach of ground forces.

5-36. After airborne forces have landed in the objective area, they reorganize and maneuver to seize objectives. When it is necessary for assault aircraft to land in the drop zone, they are parked and unloaded rapidly. Then, they may be used to transport Soldiers injured during the parachute assault. It must be understood that organic medical units may experience an overload of patients during the early phases of an airborne assault. These units have to hold the patients until either ground link-up is made or evacuation can be established at airheads. Aeromedical evacuation from the airhead is accomplished using tactical and strategic USAF aircraft.

5-37. The air assault division's organic aircraft have the ability to attack from any direction, over fly obstacles, and bypass enemy positions. Evacuation of patients in the assault phase is accomplished by AAs. Air ambulances may accompany the air assault task force (AATF) or respond from laager sites once the initial assault has taken place. If AAs are providing on-call support, it will be necessary to fly secure air avenues of approach.

5-38. When both airborne and air assault divisions have been employed and become a part of other conventional forces, their operations are similar to that of light infantry forces. During initial deployment, division medical evacuation assets may be used to evacuate patients to the airhead for air evacuation directly to division/corps hospitals.

Chapter 6

Medical Regulating

GENERAL

6-1. Medical regulating is the coordination and control of moving patients to MTFs which are best able to provide the required specialty care. This system is designed to ensure the efficient and safe movement of patients.

PURPOSE OF MEDICAL REGULATING

6-2. Medical regulating entails identifying the patients awaiting evacuation, locating the available beds, and coordinating the transportation means for movement. Careful control of patient evacuation to appropriate hospitals is necessary to—

- Effect an even distribution of cases.
- Ensure adequate beds are available for current and anticipated needs.
- Route patients requiring specialized treatment to the appropriate MTF.

6-3. The factors that influence the scheduling of patient movement include the following:

- Patient's medical condition (stabilized to withstand evacuation).
- Tactical situation.
- Availability of evacuation means.
- Locations of MTFs with special capabilities or resources.
- Current bed status of MTFs.
- Surgical backlogs.
- Number and location of patients by diagnostic category.
- Location of airfields, seaports, and other transportation hubs.
- Communications capabilities (to include radio silence procedures).

MEDICAL REGULATING TERMINOLOGY

6-4. As medical regulating may include coordination with other services, it is necessary to use the correct terminology. These terms include—

- Intracorps Medical Regulating—This is the system by which patients are transferred or evacuated from an FSB or main support battalion (MSB) to a combat support hospital (CSH).
- Intratheater Medical Regulating—This is the system by which patients are transferred or evacuated from one hospital to another within the AO. This includes evacuations between CZ hospitals, between EAC hospitals, or from CZ hospitals to EAC hospitals.
- Intertheater Medical Regulating—Moving patients between, into, and out of the different theaters of the geographic combatant commands and into the CONUS or another supporting theater. (JP 4-02.)
- Patient Administrator—The patient administrator (PAD) accomplishes the medical regulating function at the hospital level in addition to his normal duties. His medical regulating functions include consolidating all evacuation requests within the hospital and forwarding an evacuation request to his next higher headquarters for action. The PAD is also responsible for keeping his next higher MRO apprised of the current beds available and the OR status.

- Medical Regulating Officer—The MRO functions as the responsible individual at C2 headquarters for receiving and consolidating evacuation requests. These requests are initiated by the DSS or subordinate hospitals. The MRO also maintains the current patient status, bed status, and the surgical backlog at subordinate hospitals. His duties include—
  - Managing what patient classes are regulated into his facility.
  - Determining what resources are available to move the patients and coordinating for the use of these assets.
  - Maintaining accountability of patients within the MTFs.
  - Preparing reports as required.
- Theater Patient Movement Requirements Center—The TPMRC is a joint agency normally located at or near the unified theater headquarters. The theater surgeon supervises the functions of this office. These functions include—
  - Maintaining direct liaison with the Global Patient Movement Requirements Center (GPMRC), the MROs of component services, and the transportation agencies which furnish the means for evacuation.
  - Obtaining periodic reports of available beds from the services MROs providing hospitalization.
  - Selecting hospitals based on the reported bed availability to receive patients within EAC.
- Global Patient Movement Requirements Center—The GPMRC is a joint agency located in CONUS and established by the USTRANSCOM. The GPMRC receives requests from the TPMRCs. The primary role of the GPMRC is to apportion intertheater assets to the TPMRCs, collaborate and integrate proposed TPMRC intertheater plans and schedules, and communicate lift and bed requirements. The destination hospital is determined based on the patient's medical needs, the available transportation resources, and MTF capabilities.
- Joint Patient Movement Requirements Center (JPMRC)—Joint patient movement requirements center provides theater patient movement requirements center-type domain, automatic information system support and operations for a joint task force operating within a unified command area of responsibility. The theater patient movement requirements center maintains overall responsibility for theater patient movement operations, but the joint patient movement requirements center is responsible for patient movement operations within its area scope of responsibility and coordinates with the theater patient movement requirements center for intra-theater patient movement and the GPMRC for inter-theater patient movement. Also called JPMRC.
- Theater Aeromedical Evacuation System (TAES)—The TAES is a functional organization which is provided by the USAF and performs the mission of theater AE. Significant components of the TAES include the following:
  - Aeromedical evacuation command squadron.
  - Aeromedical evacuation control team (AECT).
  - Mobile aeromedical staging facility.
  - Contingency aeromedical staging facility (CASF) (under the C2 of the medical group commander).
  - Aeromedical evacuation crews.
  - Critical care air transport teams (CCATTs) (not necessarily Air Force).
- Aeromedical Evacuation Command Squadron—The AE command squadron provides C2 of all assigned TAES forces and can deploy in advance of other AE components to arrange support requirements for AE forces. The command squadron advises commanders and other personnel/agencies on AE operations, capabilities, and requirements and provides procedural and technical guidance for attached and transiting AE elements.
- Aeromedical Evacuation Control Team—The AECT is a CASF (under the C2 of the medical group commander) and is an USAF element responsible for linking validated patient movement requests with the appropriate airlift platform. The AECT is normally located within the Air

Mobility Division of the theater air operations center (AOC) for intratheater and the tanker airlift control center (TACC) for intertheater coordination of AE operations using USAF aircraft. The CONUS and strategic AOC for the USAF is the TACC located at Scott Air Force Base, Illinois. The AECT coordinates with and receives requirements from the PMRC for execution.

- Aeromedical Evacuation Liaison Team (AELT)–The AELT provides a direct communications link and immediate coordination between the user service and the TAES. The AELT is located at any level where Air Force fixed-wing requests are initiated, verifies and coordinates with the AECT/patient movement requirements center (PMRC) the physiology of flight issues and patient flight/movement requirements, assists with patient preparation for flight, and directs patient on-load activities.
- Mobile Aeromedical Staging Facility—The MASF is a USAF staging facility employed at forward airfields in the CZ to provide a temporary staging capability for preparation of patients being evacuated from corps to EAC hospitals. The MASF is employed to ensure patients are prepared for aircraft loading with the main focus of reducing AE aircraft ground time.
- Contingency Aeromedical Staging Facility—The CASF, similar to the MASF, is a USAF staging facility. The CASF, which provides greater capability and longer holding periods than the MASF, is a fixed facility located at theater AE strategic hubs and interfaces with the TAES. As opposed to the TAES elements, which are under the C2 of the AE command squadron, the CASF reports to the medical group commander.
- Military Sealift Command (MSC)—The MSC is the USN element responsible for coordinating movement of supplies, equipment, and personnel into the JOA by Navy ships. Further, it coordinates, through the TPMRC, the medical evacuation of patients by ship from the JOA to the support base or hospital ship, as required.
- Corps Movement Control Center (CMCC)—The CMCC is the corps movement control organization. It provides centralized movement control and highway regulation for movement of personnel and material into, within, and out of the corps area. When USAF capabilities are exceeded, the CMCC coordinates requests for additional air and ground resources. It also obtains the necessary clearances to support the medical evacuation mission from the CZ.
- Movement Control Battalion (MCB)—The MCB mission is to provide movement management services and highway traffic regulations and to coordinate for personnel and material movements into, within, and out of the theater. The MCB coordinates with allied and HN movement control agencies. It also coordinates with the USTRANSCOM and its subordinate units (such as the Air Mobility Command and MSC) and prepares movement and port clearance plans and programs.
- Joint Military Transportation Board (JMTB)—The JMTB is a joint staff composed of members of the Army, USAF, and USN that coordinates transportation requirements for patients requiring intertheater evacuation.
- Transportation Command Regulating and Command and Control Evacuation System—The Transportation Command Regulating and Command and Control Evacuation System is the single DOD automated information system that supports medical regulating, patient movement, and patient in-transit visibility requirements.
- Automated Patient Evacuation System—The Automated Patient Evacuation System is the system that automates the patient movement portion of medical evacuation.

MEDICAL REGULATING FOR THE DIVISION

6-5. Medical regulating in and from the division is the responsibility of the DSS. Medical regulating in the division is not as formalized as the rest of the medical regulating system. It is usually operated procedurally so as not to depend solely on communications to effect rapid evacuation. The medical regulating function in the DSS is concerned primarily with.

- Tracking the movement of patients throughout the division MTFs and into the corps facilities.
- Monitoring the use of ambulance assets.

6-6. Air and ground ambulances placed in GS of the BCTs can be field sited in the sustainment area when the mission directs. When these assets go forward to the BSMC or the ASMC to evacuate patients to corps MTFs, they have corps MTF destinations predetermined (blocks of beds). The DSS, in coordination with the medical group/brigade MRO, establishes the number of patients a supporting corps hospital can accept during a particular period of time. These blocks of available beds are then provided to the GS ambulances prior to the call for missions.

Note. Under the MRI force structure, the medical group functions were absorbed by the medical brigade.

- Once an evacuation mission is completed, the originating division MTF contacts the patient disposition section of the BSS and provides—
  - Patient numbers by category and precedence.
  - Departure times.
  - Modes of transportation.
  - Destination MTFs.
  - Any other information required by TSOP.
- The DSS, in turn, notifies the medical group/brigade MRO via the patient administration net, which is monitored by the corps MTFs. Since corps ground ambulances have no onboard communications ability and AAs have no amplitude-modulated high-frequency capability at present, all patient information is passed to the gaining MTFs via the patient administration nets. To reduce the turnaround time for ground ambulances and to move more serious patients to the CSHs in the sustainment area—
  - Air ambulances are given blocks of beds in the corps hospitals farther to the rear.
  - Ground ambulances are normally given blocks of beds in the more forward deployed CSHs.

6-7. Medical evacuation can be effected immediately, procedurally, and under conditions of communications silence without interrupting the continuum of care by—

- Preparing patient estimates.
- Prioritizing and task-organizing ambulance support.
- Assigning blocks of hospital bed designations prior to the start of the mission.
- The assistant chief of staff (personnel) (G1) is also involved in medical regulating as the G1 is responsible for personnel accountability and casualty reporting. Casualty liaison teams (CLTs) will normally be colocated at CSH to assist and coordinate patient tracking and casualty reporting.

MEDICAL REGULATING WITHIN THE COMBAT ZONE

6-8. The requirement to transfer patients from one hospital to another within the CZ occurs. This results from—

- Surgical backlogs.
- Mass casualty situations.
- Specialty care requirements.
- Planned movement of an MTF.

6-9. When it is necessary to transfer a patient, the attending physician notifies the hospital PAD. The PAD consolidates all such requests from the hospital and requests movement authority from the IPMC.

6-10. If the medical IPMC can transfer the patient or patients to its subordinate hospitals, he designates the hospitals to receive the patients and notifies both the requesting and receiving hospitals of the transfer. The IPMC also tasks subordinate medical evacuation units for the assets to transfer the patients.

6-11. If the MEDBDE cannot provide the needed hospitalization within its own resources, the MRO forwards the request to the medical TPMC for action. The TPMC then designates the receiving hospitals and notifies the subordinate IPMCs. The medical IPMCs disseminate the information to the hospital PAD and coordinate the evacuation resources for the transfer. The MRO also coordinates the regulation of patients to—

- Other US military service hospitals and naval hospital ships.
- Allied nations' military hospitals.
- Other authorized supporting facilities.

MEDICAL REGULATING FROM THE COMBAT ZONE TO ECHELONS ABOVE CORPS

6-12. Hospital attending physicians and oral and maxillofacial surgeons submit daily reports to the hospital PAD listing the patients requiring evacuation. The PAD assembles this information and transmits the report to the medical brigade headquarters. This report is a request for transportation, as well as a notification of the number of patients requiring evacuation. The report classifies the patients according to—

- Diagnostic category.
- Desired on-load points.
- When the patients will be available for evacuation.

6-13. The medical brigade TPMC consolidates these reports from each hospital attached to the medical brigade and forwards his report to the corps TPMC. The medical brigade MRO consolidates the reports and transmits the data to the theater TPMC.

6-14. If a JPMRC has been activated within the theater, the MEDCOM MRO consolidates all reports from the CZ medical brigades and forwards them to the JPMRC. The JPMRC designates hospitals in the EAC to receive the patients. The designation is based on the previously received bed status reports from all service components and available means of evacuation. The JPMRC then notifies the MEDCOM MRO of designated hospitals. The TPMC accomplishes this task if the JPMRC is not activated.

6-15. The primary means of moving patients from the CZ out of theater is USAF aircraft. With the elements of the TAES deployed, it is possible to find AELTs at each echelon and as far forward as Air Force fixed-wing requests are initiated. The PMRC monitors the MRO patient evacuation requests and passes requirements to the AELT. At the same time they pass airlift requirements to the AECT, seeking an aircraft to perform the evacuation mission. The AELT, through the MRO patient movement request, requests the PMRC/AECT to move patients. Included in the request are the originating medical facility (OMF) and the destination airfields. The airfields selected are those serving the hospitals designated to receive patients.

6-16. The AECT is a component of the TAES and performs the mission of coordinating the movement of and providing in-flight medical care to patients while under the USAF control. The AECT receives patient movement requirements from the TPMRC, then works with the airlift control team (ALCT) in the AOC to meet the evacuation requirements.

6-17. The AOC coordinates the forward movement of cargo and personnel aboard USAF aircraft with other USAF units, Army transportation representatives, and USN agencies. Certain of these aircraft are scheduled to evacuate patients on their return trips. These aircraft seldom go forward solely to evacuate patients; however, these missions may be used for retrograde airlift and mixed cargo/AE missions is the primary means of AE due to limited aircraft and the need to maximize the use of this asset.

6-18. After the schedules have been arranged, the AECT returns the detailed flight schedule to the JPMRC who passes the information to the TPMC/AELT and the parent AE element.

6-19. The MEDCOM MRO, in coordination with the AELT, issues these instructions to the IPMCs (with the authority to move patients in Army CZ facilities) and the receiving hospitals. The hospitals must prepare to receive the patients at the destination airfields. (This may be accomplished by collocating an ASMC at the airfield to receive the incoming patients.) The patients are sorted by destination hospital and moved by Army medical evacuation means. The instructions mentioned above include, as a minimum, the—

- Number of patients to be moved.
- On-load airfield.
- Destination airfield.
- United States Air Force aircraft mission number.
- Estimated time of arrival at the destination airfield.

6-20. The TPMC issues the flight and movement instructions to its subordinate medical IPMCs. The IPMCs then direct the evacuation units and hospitals within their AOs to move the patients to the on-load airfield according to the arrival time of the aircraft. This movement must be closely controlled, as a MASF can accommodate up to 50 patients. The patients cannot be delivered to the MASF earlier than 6 hours prior to arrival of the aircraft and no later than 1 hour prior to arrival.

MEDICAL REGULATING WITHIN ECHELONS ABOVE CORPS

6-21. Medical regulating within the EAC is similar to the system used within the CZ. Attending physicians within the Role 3 hospitals notify the hospital PAD of patients requiring evacuation to CSHs. The PAD then consolidates the requests from the hospital and forwards the consolidated requests to the IPMC. The IPMC, in turn, consolidates the requests and forwards them to the TPMC.

6-22. The TPMC, based on periodic bed status and availability reports from subordinate hospitals, designates specific hospitals to receive the patients. The hospitals are designated based on bed availability, to include specialty beds, to support the specific patient. The TPMC then notifies the requesting medical IPMC of the designated hospitals and, in turn, notifies the designated hospitals.

INTERTHEATER MEDICAL REGULATING

6-23. The patients who are evacuated to EAC are treated there and then further evacuated to CONUS. The attending physicians at the hospital notify the PAD. The PAD then consolidates these requests and forwards them to the IPMC. This IPMC forwards the consolidated request to the TPMC who, in turn, consolidates and forwards a request to the JPMRC (if established) or the TPMRC.

6-24. Upon request of the J/TPMRC for authority to evacuate patients to CONUS, the GPMRC directs the distribution of these patients into hospitals throughout the CONUS; advises the J/TPMRC of the destination hospital; and provides the authority for such movement. As a rule, the destination hospitals are military facilities. Civilian national disaster medical system member hospitals and other federal hospitals may also receive patients. The Department of Veterans Affairs hospitals, for example, may receive patients who are expected to be discharged from service. The GPMRC validates the TPMRC patient movement requirements and, if moving by air, tasks the TACC to plan, schedule, and execute the intertheater evacuation.

6-25. When the J/TPMRC receives the authorization to move patients, it notifies the TPMC of destination hospitals in CONUS. The TPMC coordinates with the JMTB to arrange movement of CONUS-bound patients. The TPMC then authorizes the movement to ASFs/aeromedical staging squadrons (ASTSs) that are located on or near air bases or airstrips capable of handling long-range aircraft. Transportation is arranged, within Army channels, to move patients from the hospitals to the staging facilities. The medical brigade, in coordination with the AELT, then notifies the subordinate CSHs of the flight schedule and the evacuation arrangements for movement to strategic airheads. At strategic airheads, there is an established ASF/ASTS. When the patients are delivered to the USAF, the responsibility for those patients is transferred from the Army hospital to the TAES. Upon arrival in CONUS further movement is the responsibility of the GPMRC.

6-26. All patients may not be able to be moved by air from the theater to CONUS. In that event, the MSC is used to move them by surface means. The movement authority also comes from the G/T/JPMRC or TPMC which has arranged with the Navy Service Component Command for the movement of patients by hospital ships. When the patients are moved by ships, the MEDCOM has to provide holding facilities at the port (collocating an ASMC can provide this support). Patients are delivered to these holding facilities and held there until loaded aboard the ships.

MOBILE AEROMEDICAL STAGING FACILITY

6-27. The MASF is a 13-person, mobile, tented, temporary staging facility deployed to provide supportive patient care and administration. Each MASF is capable of a throughput of 40 patients per 24 hours and should be able to sustain this tempo for 72 hours before augmentation is required and is not intended to hold patients overnight or for an extended period.

6-28. This theater system is used to evacuate patients from—

- United States Air Force operational locations within the CZ to hospital facilities outside the CZ.
- Airhead or airborne objective areas where airborne operations include USAF forward logistics support.

6-29. Mobile aeromedical staging facilities—

- Provide supportive medical care when not augmented by the CCATTs. (When augmented by CCATTs, continued stabilization of patients can be accomplished. Aeromedical evacuation crews and CCATTs fly airlift missions to provide in-flight patient care.)
- Confirm sending facility has prepared patients for evacuation.
- In coordination with the OMF, ensure patient evacuation manifests are completed. (If the OMF is an Role 2 facility [such as an FSMC augmented with an FST], the forms required to complete the manifest may not be available and the MASF will be required to complete the appropriate forms.)
- Identify patient baggage tags.

6-30. Upon deployment, the OMF provides an adequate quantity of medications for patients' transit time to the regulated destinations.

6-31. The MASF staff also establishes liaison with the OMF. The AELT is composed of one medical service corps officer for administrative assistance and a flight nurse. The team provides the initial interface between the user service and the TAES. The AELT is located at any level of the combat forces where Air Force fixed-wing requests are initiated.

LIMITATIONS OF THE UNITED STATES AIR FORCE THEATER AEROMEDICAL EVACUATION SYSTEM

6-32. There are a number of limitations that are inherent in the current system. These include the following:

- Absence of biological weapon and chemical weapon agent decontamination ability.
- The MASF does not have the capability to provide patient meals.
- The AECT ensures the initial 30-days medical resupply package arrives at the MASF. The MASF/AELT/aeromedical evacuation operations team rely on the user service for all other logistical support.
- It is the Army's responsibility to provide food and other logistical support required including moving patients back to Army facilities should USAF AE support be delayed.

ORIGINATING MEDICAL FACILITY'S RESPONSIBILITIES

6-33. Once the authorization to move the patient has been given, the OMF must complete the following administrative procedures prior to entering the patient into the TAES:

- The patient's baggage tag, patient evacuation manifest, and patient evacuation tag are the specified evacuation forms for all services and are completed as required by triservice regulation.

Note. If the OMF is a Role 2 facility (such as the BSMC), the forms may not be available; in which case, the forms will need to be completed by the MASF.

- All of the patient's medical records must be collected together and packaged. The dental records are forwarded separately in the event they are needed for identification.
- At the appropriate time, the OMF provides transportation to the MASF and assists in the offload.
- The OMF must provide the necessary medications, medical supplies, and equipment to support the patients' travel time to the regulated destination.
- Any requirements for armed guards must be met by the echelon commander.

Note. Medical units do not provide guards for prisoners or EPW in their care. When guards are required, they are provided by the echelon commander. The OMF will coordinate for this support when needed.

- A limited amount of personal baggage is authorized if each piece is properly tagged and delivered to the MASF with the patient. Patients will always be evacuated with CBRN-protective equipment, less the protective overgarment.
- Each patient must be clearly identified with a wristband or equivalent identification and properly classified as to his medical condition.
- The OMF must ensure that each patient is properly briefed and prepared for his evacuation prior to his arrival at the MASF.

MEDICAL REGULATING FOR ARMY SPECIAL OPERATIONS FORCES

6-34. As in medical evacuation, the medical regulating plan must be integrated with the ARSOF operational and logistic plan. Maximum use of opportune (operational and logistics) aircraft and command and logistics communications nets must be coordinated to expedite mission requests and ensure success.

6-35. The ARSOF medical planner must constantly coordinate with the battalion or group operations and logistics sections to obtain up-to-date information of opportune transportation assets to be used for evacuation. In an operation, or when the theater is not sufficiently developed to allow the TAES to be used effectively, the primary means of air evacuation will be those SOA or USAF SOF airframes conducting the clandestine mission. It is essential that coordination is made through the Theater Special Operations Command (TSOC) or the highest C2 element for flight medics or para rescue men to accompany the flight when backhauling the casualties. Otherwise, a medic from the SOF unit being supported may have to accompany the patient, leaving the mission without proper medical support, or the casualty may have to be transported without en route care.

6-36. For all other special operations, the supporting medical evacuation unit provides air and ground ambulances according to standard doctrinal procedures. United States Air Force MASFs or AELTs may be collocated at SOF support bases, or C2 bases, particularly during contingency operations where the build-up phase allows for prepositioning of assets.

6-37. During sustained special operations missions, the TSOC cannot afford to lose the services of low-density ARSOF skilled Soldiers who become casualties. Every effort must be made to preserve ARSOF capabilities and combat power in theater. A determination must be made as to who can be treated and returned to duty at hospitals within the EAC. As an exception to the theater evacuation policy, the combatant command may retain injured or wounded ARSOF in theater where they can be returned to limited duty without jeopardizing their recovery and health. There they can assume the support duties performed by other ARSOF Soldiers, freeing the latter for operational duties.

Appendix A

Geneva Conventions and Medical Evacuations

The conduct of armed hostilities on land is regulated by customary international law and lawmaking treaties such as The Hague and Geneva Conventions. The rights and duties set forth in the Conventions are part of the supreme Law of the Land. The US is obligated to adhere to these obligations even when an opponent does not. It is a DOD and Army policy to conduct operations in a manner consistent with these obligations. An in-depth discussion of the provisions applicable to medical units and personnel is provided in FM 4-02. This appendix discusses only those articles or actions which affect medical evacuation operations.

DISTINCTIVE MARKINGS AND CAMOUFLAGE OF MEDICAL FACILITIES AND EVACUATION PLATFORMS

This paragraph implements STANAG 2454.

A-1. All US medical facilities and units, except veterinary, display the distinctive flag of the Geneva Conventions. This flag consists of a Red Cross on a white background. It is displayed over the unit or facility and in other places as necessary to adequately identify the unit or facility as a medical facility.

AUTHORIZED EMBLEM

A-2. The Geneva Conventions authorizes the use of the following distinctive emblems on a white background: Red Cross; Red Crescent; and Red Lion and Sun. In operations conducted in countries using an emblem other than the Red Cross on a white background, US Soldiers must be made aware of the different official emblems. Although not specifically authorized as a symbol in lieu of the Red Cross, enemies of Israel in past wars have recognized the Red Star of David and have afforded it the same respect as the Red Cross. This showed compliance with the general rule that the wounded and sick must be respected and protected when they are recognized as such, even when not properly marked.

A-3. United States forces are legally entitled to only display the Red Cross. However, commanders have authorized the display of both the Red Cross and the Red Crescent to accommodate HN concerns and to ensure that confusion of emblems would not occur. Such use of the Red Crescent must be in a smaller size than the Red Cross.

This paragraph implements STANAG 2931.

Camouflage of the Distinctive Emblem

A-4. Camouflage of medical facilities (medical units, medical vehicles, and medical aircraft on the ground) is authorized when the lack of camouflage might compromise tactical operations. The marking of facilities and the use of camouflage are incompatible and should not be undertaken concurrently.

A-5. If failure to camouflage endangers or compromises tactical operations, the camouflage of medical facilities may be ordered by a NATO commander of at least brigade level or equivalent. Such an order is temporary and local in nature and is rescinded as soon as circumstances permit.

Note. There is no such thing as a *camouflaged* Red Cross. When camouflaging a medical unit or ambulance, either cover up the Red Cross or take it down. A *black cross* on an olive drab or any other background is not a symbol recognized under the Geneva Conventions.

MEDICAL AIRCRAFT

A-6. Medical aircraft exclusively employed for the removal of wounded and sick and for the transport of medical personnel and equipment shall not be attacked, but shall be respected by the belligerents, while flying at heights, times, and on routes specifically agreed upon between the belligerents concerned.

A-7. The medical aircraft shall bear, clearly marked, the distinctive emblem together with their national colors on their lower, upper, and lateral surfaces.

A-8. Unless agreed otherwise, flights over enemy or enemy-occupied territory are prohibited.

A-9. Medical aircraft shall obey every summons to land. In the event that a landing is thus imposed, the aircraft with its occupants may continue its flight after examination, if any.

A-10. In the event of involuntary landing in enemy or enemy-occupied territory, the wounded and sick, as well as the crew of the aircraft, shall be prisoners of war; medical personnel will be treated as prescribed in these Conventions.

SELF-DEFENSE AND DEFENSE OF PATIENTS

A-11. When engaging in medical evacuation operations, medical personnel are entitled to defend themselves and their patients. They are only permitted to use individual small arms.

A-12. The mounting or use of offensive weapons on dedicated medical evacuation vehicles and aircraft jeopardizes the protections afforded by the Geneva Conventions. These offensive weapons may include, but are not limited to machine guns, grenade launchers, hand grenades, and light antitank weapons.

A-13. Medical personnel are only permitted to fire in their personal defense and for the protection of the wounded and sick in their charge against marauders and other persons violating the Law of War.

ENEMY PRISONERS OF WAR/DETAINEES

A-14. Sick, injured, wounded enemy EPW, and detainees are treated and evacuated through normal medical channels but are physically segregated from US, allied, or coalition patients.

A-15. Personnel resources to guard EPW or detainee patients are provided by the echelon commander. Medical personnel do not guard EPW or detainee patients.

COMPLIANCE WITH THE GENEVA CONVENTIONS

A-16. The US is a party to the 1949 Geneva Conventions. Two of these Conventions afford protection for medical personnel, facilities, and evacuation platforms (to include aircraft on the ground). All HSS personnel should thoroughly understand the provisions of the Geneva Conventions that apply to medical activities.

CONSEQUENCES OF VIOLATIONS

A-17. Violation of these Conventions can result in the loss of protection afforded by them. Medical personnel should inform the tactical commander of the consequences of violating the provisions of these Conventions. The consequences may include—

- Medical evacuation assets subjected to attack and destruction by the enemy.
- Health service support capability degraded.
- Captured medical personnel becoming prisoners of war rather than retained persons. They may not be permitted to treat their fellow prisoners.
- Loss of protected status for medical unit, personnel, or evacuation platforms (to include aircraft on the ground).

PERCEPTION OF IMPROPRIETY

A-18. Because even the perception of impropriety can be detrimental to the mission and US interests, HSS commanders must ensure they do not give the impression of impropriety in the conduct of medical evacuation operations. For example, if a medical commander included in the unit's TSOP rules governing the use of automatic or crew-served weapons, it would give the impression that the unit possessed and intended to use these types of weapons. Under the provisions of the Geneva Conventions, medical units are only authorized individual small arms for use in the defense of the patients under their care and for themselves. Even though the unit does not possess these types of weapons, the entry in the TSOP could be misinterpreted and a case made that the commander intended to use these weapons in violation of the Geneva Conventions.

Appendix B
Legacy Units

Currently there are two units that are still active but will be transitioning out of the Army's inventory. They are the medical evacuation battalion and the medical company, AA (UH-1V).

SECTION I — MEDICAL EVACUATION BATTALION

B-1. The HHD, medical evacuation battalion, serves as the central manager of ground and air evacuation assets within the corps and EAC.

B-2. Assignment. The medical evacuation battalion is assigned to the MEDCOM in the EAC or to the medical brigade in the corps. It is normally further assigned to a medical group for C2.

Note. Under the MRI design, the medical group was replaced by the MEDBDE.

B-3. Air and ground ambulance companies assigned to the MEDCOM or MEDBDE are attached to the medical evacuation battalion for C2.

B-4. The basis of allocation is one medical evacuation battalion per a combination of three to seven of the following units:

- Medical companies, AA.
- Medical companies, ground ambulance.

B-5. Mission and capabilities.

- The mission of the medical evacuation battalion is to provide C2 of air and ground medical evacuation units within the AO. It tactically locates in the area where it can best control subordinate air and ground ambulance companies.
- The medical evacuation battalion is designed to focus on C2, planning, patient evacuation, subordinate unit support, and vehicle management. Specific capabilities are—
  - Command and control, planning and supervision of operations and training, and administration of a combination of air and ground ambulance companies.
  - Staff and technical supervision of aviation operations, safety, standardization, and aviation unit maintenance (AVUM)-level maintenance within the attached AA companies.
  - Coordination of medical evacuation operations and communications functions.
  - Coordination of logistics and service support to attached units.
  - Aviation medicine and unit-level AHS.
- This unit is dependent upon appropriate elements of the corps or ASCC for—
  - Personnel service support.
  - Health service support, to include hospitalization.
  - Mortuary affairs support.
  - Laundry, shower, and clothing repair.
  - Communications security equipment maintenance.
  - Military police support.

B-6. Organization and functions.

- Medical Evacuation Battalion (Figure B-1). The headquarters and headquarters detachment (HHD), medical evacuation battalion, is organized into a—
  - Battalion headquarters section.
  - Adjutant (US Army) section.
  - Intelligence Officer (US Army)/S3 section.
  - Supply Officer (US Army) section.
  - Detachment headquarters.
  - Treatment team.
- Battalion Headquarters Section. This section provides C2 of the assigned and attached air and ground ambulance companies. It also assists the commander on all military intelligence matters (to include the health threat), organization, training, operations, planning, personnel service support, and logistics support. Further, it provides information on the health of the command and aviation medicine expertise, as well as providing supervision over technical and flight aspects of administration, training, and safety within subordinate aviation units.
- Adjutant (US Army) Section. This section operates according to METT-TC and is the principle coordinating staff officer responsible for the delivery of human resources support. The S1 is responsible for the execution of all of the HR core competencies to include —PRM; PASR; PIM; R5 operations; casualty operations; EPS; and postal operations. It also includes MWR operations and HR planning and staff operations. The S1 also coordinates with elements of supporting agencies for finance, legal, religious, and administrative services.
- Intelligence Officer (US Army)/S3 Section. This section assists the S2/S3 officer in the execution of his duties and is capable of sustained 24-hour operations. This section remains abreast of the tactical situation and determines future medical evacuation requirements. It plans for ground and air evacuation operations, coordinates command post operations, and maintains the status of the AA units and plans for their employment. Further, this section maintains communications systems and nets, determines intelligence requirements, coordinates with movement control elements, and prepares orders and overlays.
- Supply Officer (US Army) Section. The S4 section assists the S4 officer in the execution of his duties. This section plans, coordinates, and supervises the requisitioning, receipt, storage, issue, and accounting for all classes of supply. Further, it monitors and keeps the commander informed on all matters pertaining to maintenance on assigned aircraft, ground vehicles, and medical equipment. This section also serves as the interface with the supporting medical battalion, logistics (forward/rear) for medical nonexpendable and durable item supply transactions.
- Detachment Headquarters Section. The detachment headquarters section provides, C2, administration, and logistics support for assigned personnel. It is also responsible for company supply and armament functions, food service operations, maintenance operations, and unit administration.
- Treatment Team. The treatment team provides unit-level AHS to assigned and attached elements collocated with the detachment headquarters and to adjacent units on an area support basis. The physician is a flight surgeon and provides staff assistance to the battalion commander on all matters pertaining to aviation medicine. The flight surgeon provides care and treatment for all assigned and attached aircrew members. This physician is dual-hatted as the battalion surgeon.

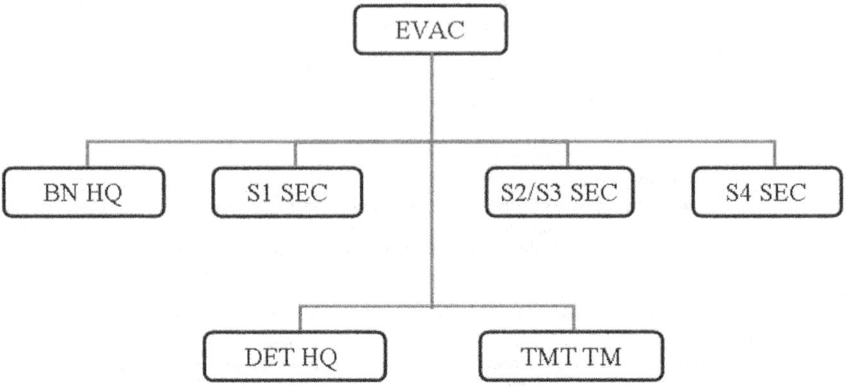

Figure B-1. Medical evacuation battalion

SECTION II — MEDICAL COMPANY, AIR AMBULANCE

B-7. The medical company, AA, provides aeromedical evacuation for all categories of patients consistent with evacuation precedence and other operational considerations. Medical evacuation is effected from as far forward as possible in the tactical AO to division- and corps-level MTFs.

B-8. Assignment.

- The medical company, AA, is normally assigned to the MEDCOM or MEDBDE and attached to the medical evacuation battalion for C2.
- The basis of allocation is one unit in support of each division or equivalent force supported. Further, one unit is in GS in the corps per two division or fraction thereof; or .333 units per separate brigades or ACRs.

B-9. Mission and capabilities.

- The mission of the medical company, AA, is to provide.
  - Aeromedical evacuation support within the theater, either DS to the divisions or GS to the corps.
  - Emergency movement of medical personnel, equipment, and supplies including whole blood, blood products, and biologicals.
- Specific capabilities of this unit are to operate on a 24-hour-a-day basis and evacuate patients based on operational capability (dependent on type of aircraft).
  - Operate 15 AAs (UH-60s). These ambulances are each capable of carrying six litter patients and one ambulatory patient, or seven ambulatory patients, or some combination thereof. Single patient lift capability is 90 litter patients, or 105 ambulatory patients, or some combination thereof. In-flight medical treatment and surveillance of patients is provided by a flight medic.
  - Operate 15 AAs (UH-1H/V). These ambulances are capable of carrying six litter, or nine ambulatory patients, or some combination thereof. Single patient lift capability is 90 litter, 135 ambulatory, or some combination thereof. In-flight medical treatment and patient surveillance are provided by a flight medic.
- Provide internal/external load capability for the movement of medical personnel and equipment.

- Perform AVUM on all organic aircraft and organizational maintenance on all organic avionics equipment. It also performs unit-level maintenance on all organic equipment less medical.
- Provide air crash rescue support, less fire suppression.
- Operate as an area support MEDEVAC section and three FSMTs to provide flexibility in supporting division, brigade, or brigade TF equivalent operations.

B-10. This unit is dependent upon:

- Support elements of corps or ASCC for—
  - Financial management, legal, and religious support.
  - Human resources support.
  - Logistics.
  - Health service support, to include medical supply and equipment.
  - Food service support.
  - Communications security equipment maintenance.
  - Mortuary affairs support.
  - Military police support.
  - Laundry, shower, and clothing repair.
  - Engineer support for heliport/landing strip construction and maintenance.
- The supporting aviation intermediate maintenance (AVIM) organization for AVIM support.

B-11. Organization and functions.

- The medical company, AA (Figure B-2), is organized into a—
  - Company headquarters.
  - Flight operations platoon consisting of a platoon headquarters, a flight operations section, and an airfield service section.
  - Aircraft maintenance platoon consisting of a platoon headquarters, a component repair section, and a maintenance section.
  - Air ambulance platoon consisting of a platoon headquarters, an area support MEDEVAC section, and three FSMTs.

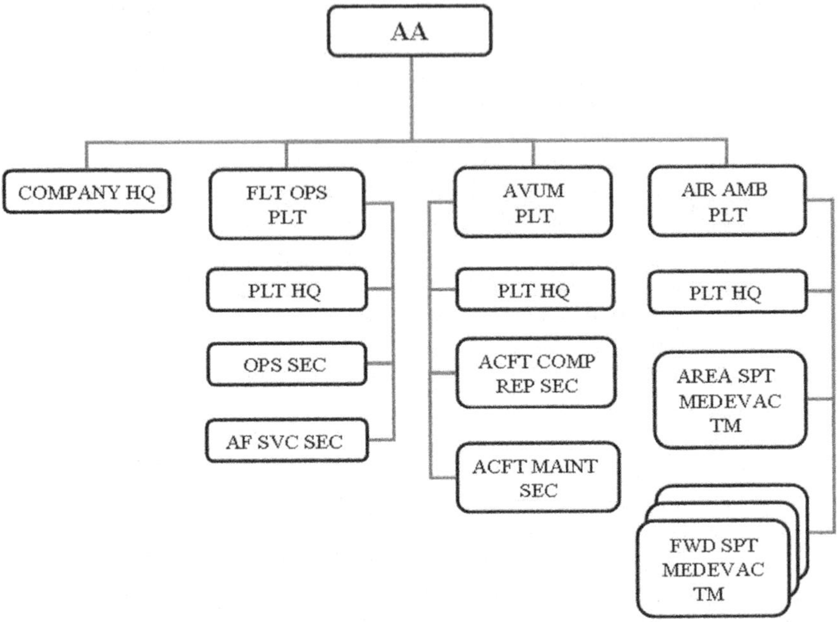

Figure B-2. Medical company, air ambulance

Appendix C

Tactical Standing Operating Procedures

This appendix provides a sample TSOP for an MMB engaged in medical evacuation missions. It should not be considered as all-inclusive. It may be supplemented with the information and procedures required for operating within a specific command or special operation.

PURPOSE

C-1. The TSOP prescribes policy, guidance, and procedures for routine support of tactical operations of a specific unit. It should cover broad areas of unit operations, but be sufficiently detailed to provide newly assigned personnel with the guidance required for them to perform their mission. A TSOP may be modified by the TSOP and OPLANs/OPORDs of higher headquarters. It applies to a specific unit and all subordinate units assigned and attached. Should a TSOP not be in conformity with the TSOP of the higher headquarters, the higher headquarters TSOP governs. The TSOP is periodically reviewed and updated as required.

FORMAT FOR THE TACTICAL STANDING OPERATING PROCEDURE

C-2. There is not a standard format for all TSOPs; however, it is recommended that a unit TSOP follow the format used by its higher headquarters. The TSOP can be divided into sections (specific functional areas or major operational areas) and further subdivided into annexes. An annex can be further subdivided into appendixes and then into tabs. Appendixes can be used to provide detailed information on major subdivisions of the annex, and tabs can be used to provide additional information (such as report formats or area layouts) addressed in the appendix.

C-3. Regardless of the format used, the TSOP follows a logical sequence in the presentation of material. It should discuss the chain of command, major functions and staff sections of the unit, operational requirements, required reports, necessary coordination with higher and subordinate elements for mission accomplishment, programs (such as command information, preventive medicine measures [PMM], and combat stress control) and other relevant topics.

C-4. Pagination of the TSOP can be accomplished by starting with page 1 and numbering the remaining pages sequentially. If the TSOP is subdivided into sections, annexes, appendixes, and tabs, a numbering system that clearly identifies the location of the page within the document can be used. Annexes are identified by letter and are listed alphabetically. Appendixes are identified by numbers and arranged sequentially within a specific annex. Tabs are identified by a letter and are listed alphabetically within a specific appendix. After numbering the initial sections using the standard numbering system (sequentially starting with page 1 through to the end of the sections), number the annexes and their subdivisions. They are numbered as the letter of the annex, the number of the appendix, the letter of the tab, and the page number. For example, page 4 of Annex D is written as "D-4"; page 2 of Appendix 3 to Annex D is written as "D-3-2"; page 5 of Tab A to Appendix 3 of Annex D is written as "D-3-A-5." This system of numbering makes the pages readily identifiable as to their place within the document as a whole.

C-5. In addition to using a numbering system to identify specific pages within the TSOP, descriptive headings should be used on all pages to identify the subordinate elements of the TSOP.

- The first page of the TSOP should be prepared on the unit's letterhead. The remaining pages of the major sections should include the unit identification in the upper right-hand corner of the page (for example: "Multifunctional Medical Battalion").
- A sample heading for an Annex is: "ANNEX B (Command Post) to Multifunctional Medical Battalion."

- A sample heading for an Appendix to Annex B is: "APPENDIX 3 (Command Post Security) to ANNEX B (Command Post) to Multifunctional Medical Battalion."
- A sample heading for a Tab to Appendix 3 to Annex B is: "Tab A (Tactical Operations Center [TOC] Security) to APPENDIX 3 (Command Post Security) to ANNEX B (Command Post) to Multifunctional Medical Battalion."

C-6. As the TSOP is developed there may be an overlap of material from one annex to another. This is due in part to similar functions that are common to two or more staff sections. Where overlaps occur, the material presented should not be contradictory. All discrepancies will be resolved prior to the authentication and publication of the TSOP. The TSOP is authenticated by the unit commander.

SAMPLE TACTICAL STANDING OPERATING PROCEDURES (SECTIONS)

C-7. The information contained in this paragraph can be supplemented. It is not intended to be an all-inclusive listing. Different commands will have unique requirements that need to be included. The first section of the TSOP identifies the specific unit/headquarters that developed it.

- Scope–This document establishes and prescribes procedures to be followed by the unit identified and its assigned, attached, or OPCON units/elements.
- Purpose–This document provides policy and guidance for routine tactical operations of this headquarters and its assigned, attached, or OPCON units.
- Applicability–Except when modified by policy guidance, TSOP, or OPLANs/OPORDs of the higher headquarters, this document applies to this unit and to all units assigned, attached, or under OPCON for combat operations. These orders, however, do not replace judgment and common sense. In cases of nonconformity, the document of the higher headquarters governs. Each subordinate element will prepare a unit TSOP, conforming to the guidance herein.
- References–This paragraph can include any pertinent regulations, policy letters, higher headquarters TSOP, or any other appropriate documents.
- General information–This paragraph discusses the required state of readiness of the unit; primary, secondary, and contingency missions; procedures for operating within another command's AO; and procedures for resolution of conflicts with governing regulations, policies, and procedures.

C-8. The second section of the TSOP discusses the specific organization.

- Organization–This paragraph furnishes specific information concerning the authority for establishing the unit, such as applicable MTOE or other staffing documentation.
- Succession of command–The guidance for determining the succession of command is discussed.
- Task organization–Task organization is contingent on the mission and will be approved by the headquarters ordering deployment.
- Joint, multinational, and interagency operations–This paragraph provides guidance on any issues concerning C2 and TF organization in joint, multinational, and interagency operations. It also provides guidance on liaison officer requirements.
- Organizational charts–Contained in Annex A.

C-9. The third section of the TSOP discusses the unit functions.

- Battalion headquarters–This paragraph discusses the C2 of the assigned, attached, and OPCON units.
- Headquarters detachment–This paragraph discusses the functions of the headquarters detachment element, such as supervising movements, internal arrangements, area layout, physical security, and operation of the headquarters and staff.
- Attached, assigned, and OPCON units–This paragraph discusses the missions and functions of these units (such as providing medical evacuation of patients, emergency movement of blood and blood products, biologicals, medical logistics, and transportation of medical personnel and equipment).

- Staff responsibilities–This paragraph lists the unit's key personnel and their duties as prescribed in FM 5-0 and any command-specific duties.

C-10. The fourth section of the TSOP pertains to staff operations and is subdivided into annexes.

SAMPLE TACTICAL STANDING OPERATING PROCEDURES (ANNEXES)

C-11. Annexes are used to provide detailed information on a particular function or area of responsibility. The commander determines the level of specificity required for the TSOP. Depending upon the complexity of the material to be presented, the annex may be further subdivided into appendixes and tabs. If the annex contains broad guidance or does not provide formats for required reports, paragraphs may be used. The annex should not require further subdivision. However, as the material presented becomes more complex, prescribes formats, or contains graphic materials, the annex may require additional subdivision. This paragraph discusses the subdivision of the annex by appendixes. It does not contain examples of subdividing the information presented in the appendixes into tabs. Applicable references such as ARs, FMs, and technical manuals should be provided in each annex. The number of annexes and their subdivisions presented below are not to be considered as an all-inclusive listing. Different commands will have unique requirements; therefore, supplementation of the information presented is permitted.

C-12. Annex A. (Organizational Charts.)

C-13. Annex B. (Command Post.)

- General. The division may operate main and/or tactical command post CPs (TAC CP) depending on the mission and tactical situation. Characteristics of the CPs include—
  - Main CP–Normally located in the sustainment area. Personnel staffing is tailored to provide planning, coordinating, and C2 of assigned/attached/OPCON units. The area location for the main CP is selected by the battalion S3; the exact site is designated by the commander in coordination with the XO and battalion S3. The XO designates work areas within the main CP. The commander, HHD, serves as the headquarters commandant. His duties include coordinating for and obtaining construction, maintenance, and logistical services and support for the CP.
  - Tactical command post CP–Normally established in the objective area or at an intermediate staging base (ISB). The ISB is the forward designated staging base at which final preparation for assault operations are controlled and is generally a safe area for support operations. The TAC CP is tailored to extend the commander's span of control and to better assess the situation on the ground and coordinate the arrival and subsequent activities of AHS organizations within the AO.
- Battalion tactical operations center.
  - Definition–The TOC is the command element of the battalion containing communications and personnel required to command, control, and coordinate HSS operations. The TOC is located within a secure, controlled area whether at a main CP or TAC CP.
  - Purpose–The purpose of the TOC is to provide the commander, in a secure environment, current evaluated information and recommendations concerning HSS/FHP operations.
  - Responsibilities–The battalion commander has overall supervision and control over the TOC. The battalion XO operates the TOC and has primary staff responsibility in the absence of the commander.
  - Operations–The TOC will operate on a 24-hour basis. It is principally staffed by each primary staff section furnishing necessary manpower as required. Secure and nonsecure telephone communications connect the TOC to other staff sections within the CP area. Access to the TOC is strictly controlled by means of an access roster, and if available, security badges. Only essential personnel and authorized visitors are allowed to enter. Each staff section will maintain an SOP on the organization and operation of its element. All elements within the TOC will, when appropriate, maintain a current situational map of their specific operations. Discussion and portrayal of tactical plans outside of the security area is prohibited.

- Composition of the TOC–This is a listing of those personnel comprising the TOC. It normally includes the commander, XO, command sergeant major, principal staff members, and other specific staff members such as the S3 (air) or the battalion aviation maintenance officer.
- Tactical operations center configuration–This is a schematic representation of the physical layout of the TOC.

- Responsibilities for the main CP services–The HHD commander's main mission is to support the headquarters, medical evacuation battalion. He has the personnel and resources assigned to facilitate this mission. He plans and coordinates for providing shower, laundry, transportation, maintenance, and other required services. He is also responsible for establishing and maintaining security; supplying fuel and other POL products; establishing the support area for the main CP (orderly room, supply room, motor pool, billets, and dining facilities); providing and operating generators to meet electrical requirements.
- Camouflage–This appendix discusses what camouflage procedures are required, to include type and amount of required camouflage materials (such as nets and shrubs); display of the Geneva Conventions distinctive emblem on facilities, vehicles, and aircraft on the ground (STANAG 2931); and other pertinent policies, guidance, or procedures.
- Message center–This establishes procedures for the handling of classified messages; provides delivery and service of IMMEDIATE and FLASH messages to the appropriate staff section; establishes procedures for preparing outgoing messages; and establishes a delivery service to the servicing message center for transmission of outgoing messages.

C-14. Annex C. (Administration and Human Resources.) This annex outlines procedures relating to administrative and human resources matters and associated activities.

- Personnel accountability and strength reporting.
  - Personnel daily summary (PDS)–This provides the procedures for filling out and submitting a daily personnel status report. The instructions may include requirements for encrypting the report prior to transmission, and specific guidance on time of submission, corrections, or other administrative requirements.
- Personnel readiness management.
  - Maintain the personnel readiness of unit personnel to include projected gains and losses.
- Personnel information management.
  - Maintain 100 percent accountability of the unit's military personnel files.
  - Update records directly through the officer record brief/enlisted record brief.
  - Reception, replacement, RTD, R&R, and redeployment.
  - Track the flow and duty status of the five categories of Soldiers into and out of the unit.
  - Update deployed theater accountability software database as required.
  - Coordinate the movement of the five categories of Soldiers.
  - Replacements–Individual replacements will not be readily available during the initial phases of operations. The S1 will automatically initiate replacement requests for personnel who are reported on the PDS report as WIA, missing in action (MIA), or killed in action (KIA).
- Essential personnel services.
  - Issue identification cards and tags.
  - Personnel actions–All personnel actions will be channeled through the S1. Company XOs and 1SGs will be the company points of contact. Actions will be handled expeditiously and meet suspense dates (tactical situation permitting).
  - Efficiency reports–This paragraph provides pertinent information on the completion and submission of these reports.
  - Award recommendations–This paragraph delineates the responsibilities for and guidance concerning submitting recommendations for awards and for scheduling and conducting award ceremonies.
  - Promotions–This paragraph discusses the procedures for submitting recommendations for promotion and scheduling and conducting promotion ceremonies.

  - Correspondence–All correspondence addressed to higher headquarters will be submitted through the S1. Requirements for submission, preparation, and approval are also provided.
- Casualty operations.
  - Casualty Feeder Card–This report is submitted on DA Form 1156 (Casualty Feeder Card). Instructions on the completion of the form and submission requirements are included.
  - Tracking of casualties will require coordination with CLT and CSH.
- Postal operations.
  - This includes hours of operation and services available.
- Morale, welfare, and recreation operations.
  - Sporting activities and morale and welfare activities.
  - American Red Cross.
  - This includes hours of operation and services available.
  - Post Exchange services–This includes hours of operation and availability.
- Management, and HR planning and staff operations.
  - Preparation of the personnel estimate and personnel portion of plans or OPORD.
  - Coordinate external HR support with HR operations cell within the sustainment brigade.
- Other administrative support.
  - Financial management–This service includes disbursements and currency control, payday activities, currency conversion, check cashing, and the appointment of Class A agents.
  - Legal services–Information and specific guidance on administrative boards, courts-martial authority and jurisdiction, legal assistance, and general services should be provided.
  - Religious activities–Religious activities include chaplain support, services available from different faiths, schedule of services, and hospital visitations.
  - Distribution–Pick up and delivery schedules and any command-specific issues/procedures are provided.
- Graves registration (GRREG)–Commanders at all levels are responsible for the recovery, identification, and evacuation of US dead. This section discusses the responsibilities and procedures for unit-level GRREG activities for assigned and attached personnel.

Note. This activity is for unit members only. Medical evacuation units do not accept nor transport the remains of nonunit members. Remains will not be transported in ambulances under any circumstances.

  - Responsibilities–This paragraph discusses both unit and battalion requirements.
  - Disposition–Specific guidance on procedures, GRREG collection points, transportation requirements, and the handling of remains are provided.
  - Hasty burials–Specific requirements for conducting hasty burials, marking, and reporting of grave sites are included.
  - Personal effects–Guidance on accounting for personal effects and requirements for burial should a hasty burial be required is contained in this paragraph.
  - Disposition of civilian and EPW remains–The local civilian government is responsible for the burial of remains of its citizens. The remains of EPW should be accomplished in separate cemeteries from US and allied personnel. If this is not possible, separate sections of the same cemetery should be used.
  - Contaminated remains–This paragraph discusses handling and disposition requirements (to include protective clothing), procedures, and marking and reporting of burial site.
- Public information–This annex contains procedures for obtaining approval on the public release of information to include the hometown news release program. In overseas locations, specific guidance on the interaction of unit members with the news media should be provided.

- Maintenance of law, order, and discipline–This appendix should provide applicable regulations, policy, and command guidance on topics such as serious incident reports, notifications, and submission formats, straggler control, confinement of military prisoners, and EPW and detainees (also discussed below).
- Enemy prisoners of war–This appendix discusses the unit responsibilities of EPW and detainees surrendered to the medical unit. These procedures do not pertain to EPW and detainee patients captured by other units. Medical personnel do not guard, search, or interrogate EPW and detainees while in the AHS; guards are provided by nonmedical personnel designated by the tactical commander for these duties. Until EPW and detainee personnel can be evacuated to an EPW collection point, medical personnel should remember and enforce the basic skills: segregate, safeguard, silence, secure, speed, and tagging. (The speed portion of evacuating EPW and detainees to designated collection points is of paramount importance to medical units.)

Note. The treatment of EPW and detainees is governed by international and US law and the provisions of the Geneva Conventions. Personnel should be aware of these requirements and have ready access to the applicable regulations and policy guidance.

- Allied, coalition, and interagency personnel–This paragraph should provide guidance on the eligibility of allied, coalition, and interagency personnel for evacuation by Army ambulance and destination locations.

C-15. Annex D. (Intelligence and Security.) This annex pertains to intelligence requirements and procedures and OPSEC considerations.

- Intelligence–The S2 has the responsibility of collecting information to assist the commander in reaching logical decisions as to the best COA to pursue. Priority intelligence requirement (PIR) may include the location, type, and strength of the air defense threat; location, type, and strength of enemy air defense radars; known or suspected CBRN activity; the health threat; and issues which the commander considers to be PIR. In addition to PIR, the commander's critical information requirements (CCIR) are also considered.
- Intelligence reports–The S2 is responsible for disseminating all applicable estimates, analyses, periodic intelligence reports, and intelligence summaries generated within the battalion or received from higher headquarters. Information on submission of reports and suspense on intelligence products and reports should also be addressed in this appendix.
- Weather data–Weather conditions can disrupt ground evacuation efforts and planners must recognize their impact when coordinating AA operations.
- Reports–These include information acquired during the routine performance of duty by pilots, ambulance drivers, and medics.
- Counterintelligence.
  - Camouflage–When ordered or directed by the tactical commander, all units will initiate and continually strive to improve camouflage operations of positions, vehicles, and equipment. Noise and light discipline is emphasized at all times.
  - Communications security–These measures will be enforced at all times. The specific requirements and considerations are included.
  - Signs and countersigns–This paragraph lists the signs and countersigns to be used during hours of darkness. It also includes reporting requirements and procedures if the sign/countersign is lost or compromised.
  - Signal operating instructions.
  - Document security–This paragraph discusses the procedures for marking and safeguarding classified material, both work documents and completed documents. (Reporting requirements in the event of compromise are also included.)
- Captured or surrendered personnel, equipment, supplies, and documents–This appendix provides specific guidance on the handling of captured personnel, equipment, supplies, and

documents. The disposition of captured medical equipment and supplies is governed by the Geneva Conventions and is protected against intentional destruction.

- Security–This appendix discusses weapons security and checks, safeguarding of patient weapons, aircraft security, signal operating instructions (communications) security, TOC security, Sensitive Item Status Report, and escape and evasion.

C-16. Annex E. (Operations.) This annex establishes procedures for S3 operations within the medical evacuation battalion, and provides a basis for standardization of medical evacuation operations in a tactical environment. It is essential that these procedures be standardized to ensure common understanding, facilitate control and responsiveness, and enhance mission accomplishment. Information on readiness levels, threat levels, warning levels, camouflage, security, area damage control, operations, and leader checklists is also included.

- Operational situation report–Requirements for preparation, format, and submission of this report are discussed in this appendix.
- Operations security–This appendix provides the guidance and procedures for secure planning and conduct of combat operations.
  - Priority intelligence requirements and security classification.
  - Responsibilities–The commander is ultimately responsible for denying information to the enemy. The S3 is responsible to the commander for the overall planning and execution of operations. The S2 has the principle staff interest in assuming the required degree of OPSEC and has the primary staff responsibility for coordinating the efforts of all other staff elements in this regards. The OPSEC officer is responsible for the preparation of the PIR and providing classification guidance. Additionally, the OPSEC officer identifies the priorities for OPSEC analysis and develops OPSEC countermeasures. The S2 coordinates with the S3 in planning OPSEC.
  - Hostile intelligence threat–The different sources of intelligence (human intelligence, signal intelligence, and so forth) are discussed.
  - Operational security program–This includes physical security, information security, and signal security.
  - Document downgrading/declassification and classification authority.
- Terrorist threat and countermeasures–This appendix provides guidance on security measures to counter the terrorist threat, both on an individual and unit level.
- Minefield threat–This appendix provides guidance on the potential of minefields being established in the AO and delineates reporting requirements. (Specific minefield extraction techniques are contained in Annex L of this TSOP.)
- Operations security and countermeasures–This appendix discusses camouflage, light discipline, noise discipline, physical security, information security, and signal security.
- Unit location update–his appendix provides timely information on the location of main and forward CPs, location of subordinate unit CPs, location of helipads, and location of POL points.
- Flight operations–This appendix provides information concerning Army aviation LZs throughout the corps area and division areas; required reports; airdrop information; and mission debriefing.
- Communications-electronics–This appendix establishes communications policies, procedures, and responsibilities for the installation, operation, and maintenance of CE equipment.
  - Responsibilities of the battalion CE noncommissioned officer in charge.
  - Concept of operations.
  - Command and control.
  - Radio communications.
  - Radio teletypewriter communications.
  - Message/communications center service.
  - Message handling procedures.
  - Wire communications.

- Switchboard operations.
- Communications security and operations.
- Intelligence security.
- Meaconing, intrusion, jamming, and interference (MIJI) reporting and electronic communications countermeasures.
- Security violations–This prescribes procedures for reporting any event of action which may have jeopardized the security of communications.
- Destruction of material.
- Daily shift inventory.
- Physical security.
- Transmission security.
- Security areas–This discusses access procedures and rosters, access approval requirements, and prohibited items.
- Inventory of classified materials.
- Communications security officers and custodians–The appointment procedures, orders requirements, and duties of personnel are described.
- Safety–This discusses requirements for grounding, handling, and storing COMSEC equipment.
- Power units.

- Rear battle responsibilities–This appendix discusses rear battle responsibilities, task organization, and support for reaction forces.

C-17. Annex F. (Chemical, Biological, Radiological, and Nuclear Operations.) This annex prescribes the policy, guidance, and procedures for CBRN defensive operations.

- Responsibilities.
- Chemical, biological, radiological, and nuclear reporting requirements and procedures.
  - Contamination avoidance.
  - Protection–Protection pertains to those measures each Soldier must take before, during, and after a CBRN attack to survive and continue the mission.
  - Decontamination–This discusses equipment requirements, procedures, and types of decontamination (such as hasty). This paragraph also provides guidance on deploying clean ambulances into a contaminated area.
  - Mission-oriented protective posture–This appendix provides guidance on the garments required for the different MOPP levels and identification procedures for personnel in MOPP.
  - Radiation exposure guidance–Establishes OEG. Discussion in this appendix includes determining what constitutes a radiological hazard, prescribing acceptable limits of potential casualty producing doses of radiation, minimizing exposure, and protecting against electromagnetic pulses.
  - Masking and unmasking procedures.
  - Radiological monitoring and survey operations.

C-18. Annex G. (Logistics.) This annex establishes logistics procedures for subordinate units when operating in a field environment.

- Nonmedical supply and services–A discussion of applicability, responsibilities, policy, classes of supply, requisition and delivery procedures, hours of operation, and other supply-relevant topics and available services (such as laundry and bath) can be addressed in this section.
- Medical logistics–This paragraph discusses supply and resupply procedures for Class VIII; policies and procedures for the backhaul of emergency Class VIII on medical vehicles; and hours of operation for the supporting medical logistics facility.
- Food service–This appendix discusses responsibilities, hours of operation, Class I supplies, sanitation requirements, layout of field kitchen, fuel storage, maintenance, safety precautions,

and administration, such as headcounts, required reports, shift schedules, meals, ready-to-eat (MRE), and inspections/visits of subordinate unit kitchens.

- Transportation/movement requirements–This appendix may cover the following areas: deployability data, applicability; responsibilities; policies on speed, vehicle markings, transporting flammable materials, transporting ammunition and weapons, and so on; convoy procedures; safety; and accident reporting.
- Fire prevention and protection–Guidance on the use of the tent stoves, flammable materials, use of cigarettes, matches, and lighters, electrical wiring and appliances, safety of tents and occupants, spacing of tents, stoves and ranges, installation of British thermal unit (BTU) heaters, and fire fighting equipment are presented in this appendix.
- Field hygiene and sanitation–This appendix provides uniform guidance and procedures for the performance of functions related to field hygiene and sanitation. It includes policies, communicable disease control, field water supply, water containers and cans, water purification bags, food sanitation, latrines, liquid waste disposal, and garbage and rubbish disposal.
- Conventional ammunition down/upload procedures–This appendix delineates responsibilities and provides guidance and procedures for the requisition, storage, and distribution of ammunition and weapons, reporting requirements, and safety.
- Petroleum, oils, and lubricants accounting.
- Medical logistics support. The medical logistics concept of operations, requisition and distribution procedures, accountability, and reports are provided in this appendix.
- Maintenance–This appendix includes information on the maintenance requirements of the battalion and the location and hours of operation of maintenance units and collection points. Maintenance for medical equipment, vehicles, aircraft, and communications and other categories of equipment are discussed.

C-19. Annex H. (Safety.) This annex establishes minimum essential safety guidance for commanders and units. It includes risk assessment, accident reporting, safety measures, emergency procedures, vehicle safety, ground guide procedures, fire prevention and protection, antennas, climate, survival training, animal and arthropods hazards, personal protective measures, hearing conservation, carbon monoxide poison, helicopter safety, and refueling operations.

C-20. Annex I. (Civil-Military Operations.) This annex discusses the participation in CMO. Medical elements are often involved in CMO, humanitarian and civic assistance, and disaster relief operations. The activities which may be covered include providing—

- Direct support medical evacuation for the operation.
- Guidance on developing a medical evacuation system in a HN.
- Training to a HN's medical infrastructure.

C-21. Annex J. (Mass Casualty Situations.) This annex discusses the procedures for providing medical evacuation support to MASCAL situations, to include coordination for nonmedical transportation assets and the augmentation of these assets with medical personnel to provide en route patient care. It discusses evacuation capabilities of US forces.

C-22. Annex K. (Minefield Operations.) This annex discusses notification procedures, training requirements, and techniques for clearing pathways to casualties, marking mines, and extracting personnel and equipment.

C-23. Annex L. (Religious Support.) This annex discusses procedures for processing unit requests for religious support and coordination of religious support to casualties, declared MASCALs, mortuary affairs support, unit memorial ceremonies/services, EPW, detainees, religious support logistics and supplies, and reports and administration.

Appendix D

Example of the Medical Evacuation Plan and of the Operations Order

SECTION I — EXAMPLE FORMAT OF THE MEDICAL EVACUATION PLAN

D-1. Figure D-1 provides an example of the medical evacuation plan.

(Classification)

Headquarters
Place
Date, time, and zone

Time Zone Used Throughout the OPLAN: *The time zone used throughout the OPLAN (including attachments) is the time zone applicable to the operation. Operations across several time zones use ZULU time. Place the classification and short title of the OPLAN at the top of the second and any subsequent pages of the base OPLAN.*

Task Organization: *List the number and coordinates of medical evacuation units here or in a trace or overlay. If you do not list units here, omit this heading.*

1. SITUATION. *(State the general medical evacuation factors affecting support of the operation. Include any information essential to understanding the current situation as it influences medical evacuation.)*

a. Enemy forces. *(Refer to the intelligence annex if it has been published or is to be published. List information about the composition, disposition, location, movements, estimated strengths, and identification of enemy forces. List enemy capabilities that could influence the medical evacuation mission.)*

(1) Strength and disposition. *(The size of the opposition force and its placement on the battlefield is important during the planning process for medical evacuation operations. When evacuation routes are selected, caution must be exercised to ensure medical evacuation assets are not compromised by going through enemy-held territory or by being ambushed by isolated pockets of resistance. Further, this information is vital in determining if medical evacuation assets [both ground and air] will require a security escort provided by movement and maneuvers forces before entering areas of the battlefield.)*

(2) Combat efficiency. *(Information on actual combat units or other forces, their training status, and their level of expertise and experience can be identified here. The level of AHS training and the development of a health care delivery system can also be included.)*

(3) Capabilities. *(This paragraph should discuss the potential capability to interfere with or disrupt medical evacuation operations.)*

(4) Logistics situation. *(This can include information on how well supplied the enemy/ opposition force is with food, clothing, or other vital logistics factors. It may also include the financial backing and availability of future support from outside individuals/groups/nations.)*

(5) Weapons. *(This should include a discussion of the enemy's weapons that present the greatest threat to air and ground evacuation personnel and vehicles. Ground assets are more likely to face a small arms threat, rocket propelled grenades (RPGs), and improvised explosive devices (IEDs) while performing their mission; air assets are vulnerable to surface-to-air weapons and when on the ground to the same threats as ground elements.)*

b. Friendly forces. *(List pertinent information concerning friendly forces that might influence the medical evacuation mission.)*

(1) Strength and disposition. *(This should include all forces [US, allied, coalition, and HN] and should be maintained on overlays. It may also include liaison officers, interpreter support, and coordination requirements.)*

(2) Combat efficiency. *(The health of the command has a significant impact on this factor. Additionally, when there are significant numbers of DNBI casualties, the medical evacuation workload will increase.)*

(3) Present and projected operations. *(Medical evacuation planners must be familiar with plans for current and projected operations. A risk management assessment of current and projected operations is conducted. Casualty collecting points and AXPs must be designated during the planning process. Medical evacuation planners must also be able to anticipate changing requirements to ensure continuous medical evacuation is available to the supported force.)*

(4) **Logistics situation.** *(Since medical evacuation vehicles conduct emergency resupply of forward deployed medical units, the medical evacuation planner must maintain visibility of the current medical logistics situation.)*

(5) **Sustainment area protection plan.** *(Medical units play an important role in the sustainment area protection plan and should be included in the planning process.)*

(6) **Weapons.** *(Medical units only have small arms for self-defense and defense of their patients. Medical vehicles and aircraft should not carry any automatic or crew-served weapons; to do so would result in the loss of the protections afforded under the Geneva Conventions for the unit, medical personnel, and patients under its care.)*

(7) **Health of the command.**

(a) **Acclimation of troops.** *(Medical evacuation crews require acclimation when introduced into an AO where the temperature range and elevation are different from their home station. Medical evacuation operations involve heavy lifting and may require extended periods of time to complete. To reduce the risk of injury to litters bearers and to facilitate the evacuation effort, medical and nonmedical personnel engaged in these operations should be acclimated to the AO and a work/rest schedule should be developed and implemented. Air ambulance crew rest requirements impact aircraft availability and must also be considered.)*

(b) **Presence of disease.** *(The presence of disease impacts medical evacuation operations in two areas. The presence of disease in the AO [usually at subclinical levels in the native population] contributes to the incidence of disease manifesting itself within the supported force. As DNBI rates increase, so do the requirements for medical evacuation. Medical evacuation personnel are also susceptible to the endemic and epidemic diseases within the AO and/or multinational force. High rates of DNBI for medical personnel will adversely impact the medical evacuation capability.)*

(c) **Status of immunizations and/or chemoprophylaxis.** *(Commanders must ensure appropriate measures are taken to protect their Soldiers from DNBI. The records of replacement personnel need to be screened to ensure all required immunizations have been received and appropriate chemoprophylaxis/barrier creams for the AO initiated/provided.)*

(d) **Status of nutrition.** *(The nutrition status of the troops involved impacts on the susceptibility to disease and environmental injuries, morale, and fatigue. The medical evacuation commander must also ensure that his evacuation crews carry sufficient supplies of MREs to sustain themselves and their patients should delays in evacuation be experienced [such as delays resulting from shifts in the battle or inclement weather].)*

(e) **Clothing and equipment.** *(Special requirements for clothing and equipment to support a particular operation should be obtained prior to the beginning of the operation [for example, cold weather clothing for extreme cold weather operations; additional blankets for patients being transported (either to add padding or warmth); or block and tackle equipment for medical evacuation operations conducted in mountainous terrain or in urban areas].)*

(f) **Fatigue.** *(Mandatory work/rest schedules and sleep plans for air crews and ground evacuation personnel must be developed and implemented.)*

(g) **Morale.**

(h) **Status of training.** *(This can include any specialized training required for the conduct of a specific operation [for example: helicopter crews require deck-landing qualifications to perform shore-to-ship evacuation with US Navy ships or orientation to the social, political, economic, religious, and ethnic issues of a HN or supported population, or training nonmedical personnel in the proper techniques for carrying litters].)*

c. **Environment.**

(1) **Terrain.** *(This paragraph should discuss any aspects of the terrain that will either hinder or enhance the execution of the evacuation mission. It should discuss both natural and man-made terrain, as medical evacuation urban areas can pose significant challenges not found on a natural battlefield. Medical personnel must be able to recover injured and wounded Soldiers from below ground and from upper levels of man-made structures. If UO are planned, armor ambulances provide added protection for medical personnel and their patients; augmentation should be requested if armor ambulances are not organic to the unit. If additional resources will be required to accomplish evacuation due to impassable or difficult terrain [such as in mountain operations where additional litter bearers and medical personnel are required], it should also be addressed here. Further, the type of terrain to be traversed, such as rugged mountain or jungle swamps may require a patient [who would otherwise be ambulatory] to be transported on a litter until easier terrain is encountered.)*

(2) **Weather.** *(This should include a discussion of current weather conditions and seasonal variants. Weather conditions impact both ground and air evacuation operations; however, the most significant impact may be on AAs as severely inclement weather can ground all aircraft. It should also discuss the impact that the weather has on the terrain [such as rivers being frozen in winter or tundra becoming impassable in spring]. The climate may pose problems with acclimation as well as place additional requirements to sustain personnel during evacuation on evacuation assets [litter evacuation in the mountains in extreme cold weather operations may require warming tents and the capability to sustain the patient during nighttime when evacuation is difficult].)*

d. **Civil considerations.** Dislocated civilian population, EPW, and detainees. *(Dislocated civilians fleeing an area of hostilities can clog roadways leading away from the conflict area. This congestion on road networks may make evacuation along these routes almost impossible. The establishment of camps to sustain these categories of personnel may interrupt the road network requiring detours and lengthened evacuation times. Injured, ill, or wounded EPW or detainees are evacuated using the same evacuation means but are segregated from US, allied, or coalition patients. Coordination for nonmedical guards for EPW/detainee* *patients being evacuated through medical channels must be accomplished with the echelon commander. A determination of eligible beneficiaries for evacuation should also be developed and disseminated.)*

2. **MISSION.** *(Statement of the medical evacuation mission.)*

3. **EXECUTION.**

a. Concept of operations. *(The concept of operations describes how the commander sees the actions of subordinate units fitting together to accomplish the mission. Commanders ensure that their concept of operations is consistent with their intent and that of the next two higher headquarters. The concept of operations describes any other details the commander considers appropriate to clarify the concept of operations and ensure unity of effort. If the integration and coordination are too lengthy for this paragraph, they are addressed in the appropriate annexes. When an operation involves two or more clearly distinct and separate phases, the concept of operations may be prepared in subparagraphs describing each phase.)*

b. Coordinating instructions. *(List only instructions applicable to two or more units and not covered in unit SOPs.)*

c. Patient estimates. *(Indicate rates and numbers by type unit/division/corps/EAC.)*

(1) Number of patients anticipated. *(The anticipated number of patients affects the number and type of medical evacuation resources required.)*

(2) Distribution within the area of operations. *(Distribution within the AO is an important consideration because varying types of terrain will generate different requirements for support. Operations conducted in mountainous terrain rely heavily on manual and litter evacuation techniques, are labor intensive, and require more time to complete. Differences in road composition, traffic density, and abundance/absence of paved roads will impact differently on the type of support required. Further, if the distance between Role 2 and Role 3 MTFs exceeds the capability of US Army AAs, the use of the USAF aircraft should be anticipated and preplanned. Coordination with the USAF is required.)*

(3) Distribution in time during the operation (evacuation time). *(As the battle progresses over time and space, medical evacuation assets may be reallocated to support the units in contact. The medical evacuation plan must be sufficiently flexible to enable the commander to shift resources as required.)*

(4) Areas of patient density. *(Areas with a high patient density may require augmentation of the medical evacuation assets supporting that location.)*

(5) Possible mass casualties. *(MASCAL situations should be anticipated when possible. Evaluation of the general threat and health threat in the AO can facilitate the medical evacuation planner in forecasting evacuation requirements. When dealing with MASCAL situations, the use of nonmedical transportation platforms should be included. Augmentation of these vehicles by medical personnel to provide en route care or CLS to provide first aid will assist in reducing the deterioration of the patient's medical condition until arrival at an MTF.)*

(6) Lines of patient drift. *(This indicates what routes injured Soldiers are most likely to take from the battlefield. It is usually the most direct route over the least demanding terrain, such as at the base of the hill rather than over the hill where climbing would be required.)*

(7) Evacuation routes/corridors. *(Evacuation routes should be preplanned, indicated on the medical evacuation overlay, and reconnaissance accomplished. Routes that provide lucrative targets, have significant obstacles to circumvent, or will be unduly congested due to fleeing refugees or other displaced persons should only be used if no other routes are available. Air corridors are provided by the supporting A2C2 element. Due the placement of hospitalization resources within the JOA, AA may be required to operate out of fixed sectors and coordination for air escorts must be coordinated.)*

(8) Support requirements. *(This paragraph discusses the type of direct support, general support, or operational control required for the operation and its command relationship. Specific guidance on trigger points and release points from a particular category of support should be included.)*

(9) Medical evacuation procedures.

(a) Evacuation overlays. *(Evacuation overlays must be developed to facilitate the evacuation effort. Both supporting and supported units must maintain and update [as required] overlays throughout the operation.)*

(b) Communications. *(Medical evacuation frequencies must be designated at the outset of the operation. Requirements for liaison officers, A2C2, and interpreters must be anticipated.)*

(c) Casualty collecting points. *(Location, staffing, and activation trigger [such as crossing a phase line] must be known to supported and supporting units.)*

(d) Ambulance exchange points. *(Location, staffing, and activation trigger [such as crossing a phase line] must be known to supported and supporting units).*

Note: If an AXP is staffed with a treatment element, it is no longer considered to be an AXP but rather it becomes the forward element of a BAS or Role 2 MTF. The AXP is only a location on the ground where the patient is transferred from one evacuation platform to another. If it is staffed, it is usually staffed by one member of the evacuation platoon to direct returning ambulances to the area where the last ambulance departed from. The element designating the AXP is responsible for staffing it. Ambulance exchange points may be rendezvous points to be used only once during the operation.

Note: Ambulance exchange points should not be used as resupply points for Class VIII. (Planning on pushing Class VIII to these points may result in the loss of the Class VIII materiel, as there may not be anyone at the point to receive the supplies.)

(e) Ambulance shuttle system. *(The ambulance [or litter] shuttle system is a management tool to facilitate the medical evacuation of forward areas.)*

(10) Casualty evacuation (transportation).

(a) Medical augmentation. *(Units and/or facilities capable of providing medical augmentation support should be identified and tasked as appropriate.)*

(b) Manual evacuation.

(c) Litter evacuation. *(A trauma specialist or CLS should be included as a member of the litter team when there is a significant distance to be covered.)*

(d) Pack animals. *(In some remote locations, the use of pack animals may be the most feasible form of transportation from the POI to a place where vehicles, aircraft, or watercraft become accessible. Whenever possible, medical personnel should accompany patients evacuated in this manner.)*

(e) Nonmedical vehicle. *(Nonmedical vehicles which could be preplanned for use for casualty transport should be identified. Further, guidance on the use of vehicles of opportunity should also be discussed. When possible nonmedical vehicles used for these purposes should be augmented with medical personnel or CLS.)*

(f) Nonmedical aircraft. *(Nonmedical aircraft which could be preplanned for use for casualty transport should be identified. Guidance on the use of aircraft of opportunity should also be discussed. When possible nonmedical vehicles used for these purposes should be augmented with medical personnel or CLS.)*

(g) Nonmedical watercraft. *(Nonmedical boats, ships, or other watercraft which could be preplanned for use for casualty transport should be identified. Further, guidance on the use of these platforms of opportunity should also be discussed. When possible nonmedical vehicles used for these purposes should be augmented with medical personnel or CLS.)*

(11) Resources available.

(a) Mode of transportation.

(b) Personnel. *(This includes a discussion on personnel strengths and the impact shortages have on the medical evacuation operations.)*

(c) Supplies and equipment. *(This includes both medical and nonmedical supplies and equipment. Litter and blanket exchanges should be covered here. If PMI are an issue, they should be discussed here.)*

(12) Use of smoke and obscurants. *(In most cases, medical units will not have a high priority for the use of smoke and obscurants; however, if the supported units have planned for its use, medical evacuation units can take advantage of the situation to clear the battlefield of casualties. Guidance on the use of colored smoke to identify the pickup location for patients can also be included.)*

d. Warfighting functions.

(1) Medical planners must take the following warfighting functions into consideration.

(a) Intelligence. (Known or suspected enemy situation. Terrain and weather. What kind of road network is in the AO? How will the predicted weather impact the traffic capability on and off the roads? Are there LZs along the axis of advance and in or around the urban area? How will the weather impact on the availability of aircraft?)

(b) Movement and maneuver. *(What is the combat scheme of maneuver and are there medical assets readily available to care for potential casualties? Particular attention should be paid to the employment of the armored ambulances.)*

(c) Fire support. *(Are there any buildings specifically targeted? If so, medical planners should ensure CCPs, evacuation routes, and treatment elements remain at a safe distance away from these buildings to prevent potential fratricide incidents.)*

(d) Protection. *(Preserving the force includes protecting personnel (combatant and noncombatant), physical assets, and information of the US and multinational partners.)*

(e) Sustainment. *(Typically, the BCT moves a forward logistics element toward the battle area to provide support. If this is the case, then how are medical assets integrated into the plan, and, more importantly, where are they positioned in the convoy and how are they secured?)*

(f) Command and control. *(The senior military on-site or the senior medical person at the scene, if available, assigns the evacuation precedence and determines if the patient should be evacuated by ground or air resources. What are the primary and secondary frequencies? This element is the central point for prioritizing assets and medical evacuation requests. It is imperative that this element is capable of continuous situational awareness through accurate battle tracking.) From information provided by the S2, medical planners must understand the tactics of the enemy and be prepared for casualties at all possible areas of contact. The medical planners should be able to gain insight into this matter during the war gaming process. The medical planners should provide input into the counterreaction portion of the action, reaction, counterreaction drill.)*

(2) Additional planning information.

(a) *Availability and numbers of ground and AA assets, evacuation routes and flight windows.*

(b) *Nonstandard ground and aircraft availability (for MASCAL situations) and how to trigger their involvement.*

(c) *Additional medical assets from higher levels such as treatment teams, ambulances, and combat configured loads of Class VIII.*

(d) *Existing and proposed locations of all available MTFs.*

(e) *Class VIII resupply plan.*

(f) *The need for litter bearer teams and special equipment for extracting casualties from buildings and rubble.*

e. Casualty evacuation synchronization.

(1) *Medical evacuation and CASEVAC assets must be incorporated into the evacuation plan. Ideally there should be a single dedicated controlling authority. This prevents duplication of effort and confusion and enables the battlefield to be cleared faster and more efficiently.*

(2) *In MASCAL situations CASEVAC assets once put into operation should be dedicated versus designated (lifts of opportunity). Potential problems that arise arising from relying on designated versus dedicated CASEVAC include excessive time spent acquiring the information needed to transport casualties. Medical regulating to higher roles of care facilities may have to be accomplished by blocks of beds rather than on a by name basis. If a facility is designated for certain specialty care [such as head trauma] the evacuator must know which facilities have been identified and where they are located.).*

(3) *Another consideration is aircraft configuration. Casualty evacuation may be a seats-out operation. By dedicating aircraft and crews, they have time to prepare for the mission. Casualty evacuation crews must be briefed on the details of their mission including LZs, PZs, air corridors, communications architecture, and supported units.*

(4) *Any CASEVAC plan should clearly articulate the following information.*

(a) *Who is responsible (such as the company 1SG) for coordinating the company CASEVAC mission?*

(b) *Combat lifesavers role.*

(c) *Class VIII and blood resupply procedures.*

(d) *Personnel to act as litter bearers.*

(e) *Locations of CCPs, forward treatment teams, BAS, and AXPs.*

(f) *Location and designation of MTFs.*

(g) *Medical regulating officer coordination.*

(h) *Medical treatment facility and MRO contact frequencies.*

4. SERVICE SUPPORT.

a. Supply. *(Refer to TSOP or another annex whenever practical.)*

(1) General supply. *(Provide special instructions applicable to medical evacuation units.)*

(2) *Consider supply levels for all classes of supply, in the event of mission requirements in an austere environment and at extended distances from the full compliment of logistics and sustainment resources.*

b. Medical logistics. *(Provide special procedures applicable to the operations.)*

(1) Requirements. *(For sustaining the ground and AA operations and personnel.)*

(2) Distribution. *(This should include the method of distribution and any limitations or restrictions that are applicable. Additionally, if special transportation requirements exist, they should also be noted.)*

(3) Medical logistic activities. *(This includes the location of the medical supply activity supporting the AO and the means of communicating requests for resupply.)*

c. Supplies required for ground and AA operations and personnel. *(Includes estimates of the number of patients anticipated to be evacuated.)*

d. Health service support. *(Include information and instructions for supported units. Prescribe the plan for ground and air medical evacuation.)*

(1) Evacuation. *(List CCPs and AXPs. Establish ambulances shuttles, routes, means, and schedules of evacuation and en route treatment policies for the use of nonmedical transportation assets; policies for evacuation by ground and air; and evacuation of CBRN-contaminated patients. Include information about medical evacuation request procedures and channels, and holding policies.)*

(2) Treatment.

(a) Treatment of supported forces.

(b) Policies. *(Should address treatment and hospitalization policies.)*

(c) Units. *(This includes information concerning the location, capabilities, and communications means of units providing support.)*

(d) Other. *(This can be information on other medical assets that may be used in emergencies.)*

e. Personnel.

(1) Current subordinate unit manning levels/critical MOS shortages (consider all manning influences—Task Organization, R&R flow, Boots On Ground data, and so forth).

(2) New personnel requirements resulting from the operation (language skills, additional skill identifier [ASI], and so forth).

(3) Casualty estimates (as developed).

(4) Forecasted replacement availability.

(5) Evacuation policy for the operation.

(6) Supporting HR organizations (location, command/support relationship, controlling element).

(7) Supporting medical elements

(8) Replacement fill priority (Coordinate w/S3/G3).

(9) Crew/key leader replacement.

(10) Projected postal flow/limitations.

(11) Change to established PASR reporting flow/times.

(12) Change to casualty reporting flow (CLT change, reporting changes).

(13) RSOI Reception HR impacts.

(14) R&R schedule/flow operational impact (during operations conducted during sustained operations).

(15) Civilian and JIM manning impacts on the operation focusing on strengths and personnel service requirements.

(16) Specific Army G1 personnel planning guidance impacts on the operation.

(17) Status of other personnel service or personnel support factors (postal, MWR, legal support, military pay support).

f. Coordinating instructions. *(List instructions applicable to two or more units and not routinely covered in unit SOPs. This is always the last subparagraph in paragraph 4. Place complex instruction in an annex.)*

(1) Time or condition when a plan becomes effective. *(Include the time or the conditions under which the plan is to be placed in effect.)*

(2) Commander's critical information requirements. *(List only in coordinating instructions and not in annexes.)*

(3) Risk mitigation control measures. *(These are measures unique to this operation and not included in unit SOPs. They must include MOPP, OEG, troop safety, vehicle recognition signals, and fratricide prevention measures.)*

(4) Rules of engagement. *(If required.)*

(5) Boundaries. *(Address measures established to protect units. Specify which tactical units are to provide protection.)*

g. Special reports. *(List reports requiring special emphasis that is not addressed elsewhere.)*

5. **COMMAND AND SIGNAL.**

a. Command. *(State the map coordinates for the CP locations and at least one future location for each CP. Identify the chain of command if not addressed in unit SOPs.)*

b. Signal. *(Include the headquarters location and movements, liaison arrangements, recognition and identification instructions, and general rules concerning the use of communications and other equipment, if necessary. Use an annex when appropriate.)*

(Classification)

Figure D-1. Example of a medical evacuation plan

SECTION II — EXAMPLE FORMAT FOR AN OPERATIONS ORDER

D-2. Figure D-2 provides an example of the medical operations order.

(Classification)

Place the classification at the top and bottom of every page of the OPLAN.

(Change from verbal orders, if any)

The phrases "No change from verbal orders" or "No change from verbal orders except paragraph #" are required. This statement is only applicable if the commander issues a verbal order.

Copy ## of ## copies
Issuing headquarters
Place of issue

Show the name of the town or place in capital letters, coordinates in parentheses, and the country in capital letters. You may encode both.

Date-time group of signature

(The effective time for implementing the plan or order is the same as the date-time group of signature unless the coordinating instructions state otherwise. Use time zone Z unless the order states otherwise. When orders apply to units in different time zones, use Z time. When an OPORD does not specify the actual time to begin an operation, state that time in terms of an event.)

OPERATION ORDER 00-001, Operation Liberation
References: OPORD 00-000
Time zone: Zulu
Task Organization

Ground Ambulance Units participating in Operation Liberation
123 MED CO (GA) assigned to 1st MMB and provides emergency support to the 1st Marine Expeditionary Force (MEF)
234 MED CO (GA) assigned to 1st MMB
Ambulance Platoons assigned to:
C CO BSMC
C CO BSMC
C CO BSMC
C CO BSMC

Air Ambulance Units participating in Operation Liberation
C 2-1 MED CO (AA) CAB assigned to 1st Infantry Division
C 2-2 MED CO (AA) CAB assigned to 2nd Infantry Division
C 2-3 MED CO (AA) DS to the 1st MEF
C 2-4 MED CO (AA) CAB assigned to 1st Corps support

Air Force aeromedical evacuation assets
2d AF Combat Search and Rescue (CSAR), 1st AF Squad

1. **SITUATION.**

a. **Enemy forces.** *As addressed in OPORD 00-000, the ground-to-air threat is significant in the all enemy sectors. The air–to-air threat is minimal, in all sectors. All ground planned evacuation routes will be targets for ambush by enemy forces and sympathetic host nationals.*

b. **Friendly forces.** *A six (6) aircraft (HH-60) Air Force CSAR unit will be staged from the APOD. They will respond to downed aircrew evacuation recovery requests from the Air Force Combat Control Center.*

c. **Attachments and detachments.** *None.*

d. **Assumptions.** *None.*

2. **MISSION.** *On order provide tactical and operational ground and air intratheater medical evacuation or patient movement in support of Operation Liberation and continue coverage of all combined joint task force (CJTF) assets until relieved.*

3. **EXECUTION. Intent: State the commander's intent.**

a. **Concept of support operations.** *This is a three phase operation.*

Phase 1—Will be prior to initiation of the attack. *All US forces will be in the staging areas and in preattack positions. There will be no or very limited direct contact with enemy forces. C 2-1, C 2-2, and C 2-3 will not be required to* *provide evacuation support until D-3 days. Units in the staging area will be directed to call C 2-4 for evacuation at all times prior to D-3 days. All other evacuation units will concentrate on training and preparation for hostilities as directed by their commanders.*

Phase 2—Will be the attack. *C 2-1, C 2-2, C 2-3 will provide DS to their assigned, GS to the corps sustainment area and the backhaul mission.*

Phase 3—Will be from the conclusion of the attack until stability operations commence. *Evacuation assets will begin transferring to an area support mission during this phase.*

Note. The concept of support operations takes into account both the tactical OPORD and the medical Annex to the OPORD. Information in both documents will shape the evacuation plan. Simultaneous planning should therefore occur.

---

b. Tasks to maneuver units: *Not applicable.*

c. Tasks to movement and maneuvers units:

(1) Evacuation responsibilities.

(a) Phase 1.

C 2-1, C 2-2, C 2-3 will provide direct support to their division assets from D-3 on. C 2-1, C 2-2, C 2-3 will respond to any request for evacuation that is geographically close to their positions even if not in their division area.

C 2-4 will provide support to the entire staging area (AO Buffalo). C 2-4 will perform the ship-to - shore mission as directed by the 11th MEDCOM TPMC

The 123 MED CO (GA) will operate in the 1st MEF sector in a direct support role. 1st MMB has will cover the 1st and 2nd Division footprint.

The 234th MED CO (GA) will provide support to Camps Jones, Smith, and Franks and perform patient transfers from camps to the APOD.

(b) Phase 2.

C 2-1, C 2-2, C 2-3 will provide direct support to their division assets. C 2-1, C 2-2, C 2-3 will be responsible for moving patients from within the division area back to the staging AO Buffalo until FLOT is beyond phase line alpha.

After FLOT is beyond phase line alpha (at which time one way flight time for the H-60 back to AO Buffalo will exceed 2 hours), C 2-1 will designate 1 FSMT to stage at XYZ Airfield to perform the backhaul and the area support mission until relieved. United States AF aeromedical transports will be coordinated for by 11th MEDCOM flying from XYZ airfield.

Beyond phase line Alpha C 2-2 will provide one FSMT to support the greater casualties expected in C 2-2 sector. Coordinate internally for link-up.

C 2-4 will provide support to the entire staging area (AO Buffalo). C 2-4 will perform the ship-to-shore mission as directed by the 11th MEDCOM MRO. C 2-4 will designate 4 aircraft to go forward as required to backhaul patients to AO Buffalo.

(c) Phase 3.

C 2-1, C 2-2, C 2-3 will provide direct support to their Divisions' assets.

The 1st MEF will identify areas of high casualty density conducive to ground evacuation (urban areas) and divide the resources of the 123rd MED CO (GA) to cover these areas efficiently, if required.

The 234th MED CO (GA) will conduct convoy support and provide coverage for AO Buffalo.

d. Evacuation mission guidelines.

(1) Phase 1, 2, and 3. *The senior military person advised by medical personnel on-scene will assign the evacuation precedence and determine the mode of evacuation (ground or air). Launch authority, a separate issue, based on risk will be according to the aviation SOP.*

(a) *Weather below acceptable minimums.*

(b) *No acceptable LZ in the immediate vicinity, except for hoist missions. Ground evacuation may be used for part of the evacuation.*

(c) *Chemical, biological, radiological and nuclear conditions at LZ would contaminate the aircraft. Ground evacuation may be used for part of the evacuation, followed by an air pickup after decontamination is complete.*

(d) *Patients are located at a LZ in direct enemy contact. Ground evacuation may be used for part of the evacuation, followed by an air pickup once clear of the enemy.*

(e) *Air evacuation assets are unavailable.*

(f) *Mission flight route requires an armed escort and one is not available.*

(2) The following are additional guidelines for evacuation:

(a) *Ground evacuation is the preferred method to clear casualties out of hostile urban areas. Air ambulance pick-up at a secured LZ outside of the urban area is preferred. For large urban areas, this may not be possible.*

(b) *If a mission requires the use of a hoist (such as thick forest or jungle) and the pick-up area has suspected enemy contact, ground evacuation is advised to a secure pick-up site.*

(c) *Prescheduled movement of ROUTINE or CONVENIENCE patients along secure routes of evacuation should be conducted using ground evacuation if the distance of evacuation is short.*

(3) Coordinating instructions:

(a) Medical regulating. *The 11th MEDCOM will oversee medical regulating during the operation.*

(b) Tentative locations of MTFs.

Phase 1

- *2nd CSH AB 1234 5678*

- *1st CSH AB 2345 6789*

Phase 2

- *2nd CSH AB 1234 5678*
- *1st CSH AB 2345 6789*

Phase 3

- *2nd CSH AB 1234 5678*
- *1st CSH AB 2345 6789*

(c) Reporting. *Evacuation units, through division G3 if assigned to a division, will submit a daily report of evacuation missions by 1200 the following day.*

(d) *If a MASCAL develops which overwhelms the resources of a single AA CO, contact GSAB for assistance.*

(e) Casualty collection. *Military treatment facilities can call 9 line missions to any medical evacuation asset for URGENT and PRIORITY missions however MTFs that need to clear beds will coordinate mass patient movements through CJTF G3 air for possible USAF support.*

- *1st CSH will call C 2-4 for evacuation requests during Phase 1.*
- *2nd CSH will call C 2-4 for evacuation requests during Phase 1.*
- *1st CSH will call C 2-1 for evacuation requests during Phase 2/3.*
- *2nd CSH will call C 2-2 for evacuation requests during Phase 2/3.*

(f) Ground evacuation. *Seasonal monsoons are expected to washout many of the main planned MSRs and cause flooding.*

(g) Air evacuation. *If air evacuation routes become so long as to seriously degrade the capabilities of an AA company, contact CJTF G3 Air for additional support and possible reallocation of evacuation assets. An AA CO is considered to be operating beyond its capability or inefficiently if:*

- *More than 50 per cent of missions require a one way trip time exceeding 2 hours.*
- *On more than 50 per cent of missions flown, routes of evacuation are flown through another AA company's area of direct support coverage.*

(h) Medical considerations for evacuation.

- US Army evacuation assets will be used to evacuate all injured personnel on the battlefield without prejudice.
- Human immunodeficiency virus is at epidemic proportions in Simula population. Take appropriate precautions during treatment and when dealing with biologicals.
- A significant number EPW and displaced civilians and detainees are predicted to be evacuated. Estimated to be a much as 75 per cent of total patient load.

(i) United States Air Force aeromedical evacuation.

- Combined joint task force G3 Air will coordinate for USAF intratheater aeromedical missions.

4. SERVICE SUPPORT.

a. Casualty estimates. Reference the Medical Annex for specific casualty estimates.

b. Aviation maintenance support. C 2-3 will receive AVUM/AVIM support from the 2-4 CAB.

c. Divisions G3 will report the following CCIR to CJTF G3 Air. Use the CJTF CCIR format.

- Any MASCAL.
- Any evacuation company with degraded capability for—
- Evacuation distance(defined above)
- Maintenance: below 70 percent operational readiness rate
- Personnel: number of crews less than available aircraft
- Any aircraft Class A accidents/events or shoot downs.

5. COMMAND AND SIGNAL.

a. Fragmentary orders (FRAGOs) to this order will be published to adjust the corps evacuation plan as necessary.

b. 11th MEDCOM MIRC Chat net name, Phone xxx-xxxx

- 1st Division Surgeon (G3), MIRC Chat net name
- 2-1 CAB MEDEVAC Cell, MIRC Chat net name
- 2nd Division Surgeon (G3), MIRC Chat net name
- 2-2 CAB MEDEVAC Cell, MIRC Chat net name
- 1st MEF, MEDEVAC Cell, MIRC Chat net name
- 1st Corps CAB MEDEVAC cell, MIRC Chat net name, phone xxx-xxxx

ACKNOWLEDGE:
PATRICK, LTG

Figure D-2. Example of a medical operations order

Appendix E

Use of Smoke and Obscurants in Medical Evacuation Operations

This appendix provides guidance for AMEDD personnel in the use of smoke and obscurants in medical evacuation operations.

TACTICAL COMMANDER

E-1. Throughout the battlefield, forces acquire and engage targets based on visual, laser, and microwave technologies. Friendly and enemy units use smoke and obscurants across the battlefield as a combat multiplier. The use of smoke and obscurants to mask combat operations is dictated by the tactical commander. He normally provides the operational guidance for units or elements operating in an area requiring obscuration. Permission to employ smoke and obscurants solely to mask medical evacuation operations may not be approved. However, if the tactical commander's plan indicates that smoke operations are to be employed in a given AO, the medical planner should consider both the advantages and disadvantages posed by their employment.

FACTORS TO CONSIDER

E-2. The medical planner should consider the factors which might impact the use of smoke and obscurants in medical evacuation operations. Factors to consider are the—

- Phase of the tactical operation in which smoke and obscurants will be employed.
- Effect on ground and air evacuation routes when operating in an obscured environment (such as limited hours of use, checkpoint or convoy requirements, or the elimination of NOE approaches).
- Potential for exploiting the use of the cover and concealment provided for clearing the battlefield of casualties, especially in defensive operations.
- Potential requirements for smoke generation to perform the medical evacuation mission which would not detract from the tactical capability and requirements.

E-3. Smoke can also be used to identify unit areas or LZs for which a medical evacuation request has been received. Further, smoke can indicate wind direction at a landing site for AA operations.

EMPLOYMENT OF SMOKE AND OBSCURANTS

E-4. Smoke and obscurants are employed to protect friendly forces from attack during the offense or defense. Smoke and obscurants disrupt enemy combat operations throughout the depth of the battlefield and across the operational continuum. They—

- Disrupt the ability to communicate.
- Conceal friendly forces.
- Deceive the enemy.
- Identify and signal.
- Degrade the effects of directed-energy weapons.

E-5. The benefit to AHS forces is derived through the tactical commander's use of smoke to obscure friendly tactical maneuvers. This obscuration—

- Prohibits the enemy from knowing how many casualties have been inflicted.
- Aids the movement of medical units and equipment.

- Enhances the ability to resupply forward deployed AHS elements.
- Aids in the tactical deception plan.

GENEVA CONVENTIONS AND THE USE OF SMOKE AND OBSCURANTS IN MEDICAL EVACUATION OPERATIONS

E-6. As discussed above, the 1949 Geneva Conventions for the Amelioration of the Condition of the Wounded and Sick in Armed Forces in the Field (GWS) provides protection of medical personnel and units from intentional attack so long as they carry out no duties harmful to the enemy (Article 21, GWS). In order to facilitate their identification so as to prevent their intentional attack, medical units, equipment, and personnel are authorized to display the distinctive emblem of the Red Cross (Article 41 and Article 42, GWS). Under tactical conditions, when requirements for concealment outweigh those for recognition, all distinctive emblems may be obscured or removed from medical equipment if ordered by competent military authority and authorized by Army regulations. Display of the distinctive emblem is not required to afford the right against intentional attack; attack of medical units, equipment, and personnel not displaying the distinctive emblem is prohibited if opposing forces realize that the forces about to be attacked are medical units performing humanitarian duties.

E-7. The use of smoke or obscurants in medical evacuation operations does not differ from the use of camouflage techniques and is not prohibited by the GWS. Its only effect will be to obscure the identity of units as they perform their humanitarian mission. Given the lethality of the modern battlefield, however, it would be difficult, if not impossible, to say that such obscuration of these units, equipment, and personnel would necessarily increase their risk from unintentional attack.

E-8. It is recognized that, with the advent of precision-guided munitions and electro-optical or laser target acquisition devices, there will be a substantial use of smoke and other obscurants on the modern battlefield as a result of normal combat operations. The legitimate use of obscurants by combatants to thwart the accuracy of precision guided munitions may increase the risk to units and equipment not employing obscurants. This may possibly place medical units and equipment at greater risk if they fail to employ them. Further, medical evacuation operations will have to be carried out on the battlefield as medical personnel find it, which will include obscurants employed for normal combat operations.

USE OF SMOKE IN AEROMEDICAL EVACUATION AND HOIST RESCUE OPERATIONS

EFFECTIVE USE

E-9. Smoke can be used effectively in aeromedical evacuation and overland hoist rescue operations to—

- Identify the landing site. Colored smoke is an excellent daytime marking method. The smoke generated from a smoke grenade is difficult to detect more than 2 to 3 miles away, but an aircraft in the area should have little difficulty in noting its location. As more than one unit may be operating in a given area, it is important that the unit requesting an aeromedical evacuation mission be able to signal the aircraft as to the correct landing site to use. Radio communications produce an electronic signature. The electronic signature created from a prolonged transmission to guide an AA to the landing site may not be an acceptable tactical risk.
- Ensure the LZ is controlled by friendly forces. When a unit employs colored smoke to mark a landing site, the aircrew should identify the color and confirm it with the ground personnel. The transmission time required for this procedure is limited, thereby reducing the electronic signature.
- Determine surface wind direction. The employment of smoke at the landing site also enables the aircrew to determine the wind direction.
- Provide cover and concealment. In some environmental conditions (such as desert operations), the phenomenon of inversion occurs. When this occurs, the smoke and obscurants used in

normal combat operations may provide an upper layer of smoke under which the AA can operate.

Disadvantage

E-10. The use of smoke on aeromedical evacuation operations can be a disadvantage if incorrectly employed or if the smoke generated in the tactical operation interferes with the medical evacuation mission. Smoke can obscure the landing site and make NOE approaches unusable. Further, smoke on the battlefield can force aircraft to fly at higher than planned heights. This increases the risk of being acquired by the enemy.

Overwater Hoist Operations

E-11. In overwater hoist rescue operations, the employment of smoke for marking the patient pickup area, for determining surface wind conditions, and for spatial orientation is essential. The smoke employed by the aircrew must not interfere with the conduct of the operation or mask the location of the individual to be rescued.

EMPLOYMENT OF SMOKE IN GROUND MEDICAL EVACUATION OPERATIONS

E-12. The employment of smoke during ground evacuation operations must be in consonance with the tactical commander's plan. Smoke can mask medical evacuation operations on the battlefield, but must not interfere with the tactical mission. In all combat operations, but especially in UO, smoke can be employed to cover and conceal—

- Movement across open areas.
- Extraction of casualties from vehicles and/or buildings.
- Entry and exit into/out of structures.

Appendix F

Evacuation Capabilities of United States Forces

This appendix provides the evacuation capabilities of US forces and aircraft.

EVACUATION CAPABILITIES OF UNITED STATES AIR FORCE AIRCRAFT

F-1. Evacuation capabilities of USAF aircraft are provided in Table F-1. As the majority of USAF aircraft are not dedicated medical evacuation platforms, when used for medical evacuation the crew must be augmented with medical personnel to provide in-flight care.

F-2. The evacuation capability of Civil Reserve Air Fleet aircraft is provided in Table F-2.

Table F-1. Evacuation capabilities of United States Air Force aircraft

Transport Aircraft	*Litter*	*Ambulatory*	*Combination*
C-130 Hercules	74	92	Variety
C-141 Starlifter	48	170	Variety
C-5 Galaxy	70		
C-17A	*36	*54	
KC-135 and KC-10	8	24	
*Normal configuration			

Table F-2. Civil Reserve Air Fleet capabilities

Transport Aircraft	*Litter*	*Ambulatory*	*Combination*
Boeing B-767	*111	22	87 Litter 22 Ambulatory
* When activated, these aircraft are preconfigured for medical evacuation. The crew must be augmented with medical personnel			

EVACUATION CAPABILITIES OF UNITED STATES ARMY VEHICLES AND AIRCRAFT

F-3. The evacuation capabilities of US Army vehicles and aircraft are provided in Table F-3.

Table F-3. Evacuation capabilities of United States Army vehicles and aircraft

Vehicle/Aircraft	*Litter*	*Ambulatory*	*Combination*
Ground Vehicles			
M996 Truck, Ambulance	2	6	1 Litter 3 Ambulatory
M997 Truck, Ambulance	4	8	2 Litter 4 Ambulatory

M1010 Truck, Ambulance	4	8	
M792 Truck, Ambulance	3	6	2 Litter 3 Ambulatory
M1133, Stryker MEV	4	6	
Bus, Motor, 44 Passenger	18	37	
M113 Carrier, Personnel	4**	10	
M998 Truck, Cargo/Troop Carrier (Four Man)	3	4	
M998 Truck, Cargo/Troop Carrier (Two Man)	5		
Truck Cargo, 2 1/2 Ton, 5 Ton	12	16	
M997 Heavy Expanded Mobility Tactical Truck	9		
M871 Semitrailer, Cargo	16		
M1085 Truck, Cargo, Medium Tactical Vehicle, Long Wheel Base, 5 Ton	12	22	
M1093 Truck, Cargo, Medium Tactical Vehicle Air Drop/Air Delivery, 5 Ton	8	14	
M1081 Truck, Cargo, Medium Tactical Vehicle Light Air Drop/Air Delivery, 2 1/2 Ton	7	12	
Rotary-Wing Aircraft			
***UH-60 Blackhawk	6	7	*4 Litter 1 Ambulatory
****UH-60A Blackhawk	3	4	4 Litter 1 Ambulatory
UH-1H/V Iroquois	6	9	*3 Litter 4 Ambulatory
CH-47 Chinook	24	31	#
Fixed-Wing Aircraft			
C-12 Huron		8	

LEGEND
* Normal configuration
** Spall liner must be removed
*** without internal hoist installed, requires 6 man litter configuration kit
**** With internal hoist installed
# See Table F-5

Railway Car Capabilities

F-4. Although railway cars are not available within the US Army inventory, it is important to know the approximate capacities of rail transport in the event they become available and/or required in civil support operations or through wartime HN support agreements. The approximate capabilities of railway cars are provided in Table F-4.

Table F-4. Capabilities of railway cars

Rail Transport	*Litter*	*Ambulatory*	*Combination*
Pullman Car (US)	32	48	
Sleeping Car (NATO/HN)	32	32	
Ambulance, Railway Car (NATO/HN)	24	30	
Ambulance, Railway Car, Personnel	21	21	
Railbus			40 Litter 16 Ambulatory

EVACUATION CAPABILITIES OF UNITED STATES NAVY SHIPS, WATERCRAFT, AND ROTARY-WING AIRCRAFT

F-5. The evacuation capabilities of USN ships, watercraft, and aircraft are provided in Table F-5. The entries for ambulatory and litter patients on ships are the same because all patients require a bunk.

Table F-5. Evacuation capabilities of United States Navy ships, watercraft, and aircraft

Ships/Watercraft/Aircraft	*Litter*	*Ambulatory*
Hospital Ships		
T-AH 19, US Naval Ship Mercy	1000	1000
T-AH 20, US Naval Ship Comfort	1000	1000
Casualty Receiving And Treatment Ships		
Amphibious Assault Ship (LHD)(Multipurpose)	604	604
Amphibious Assault Ship (LHA)(General Purpose)	367	367
Amphibious Assault Ship (LPH)	222	222
Amphibious Transport Dock (LPD)	14	14
Dock Landing Ship (LSD)	108	108
Limited Medical Capability Watercraft		
Amphibious Cargo Ship (LKA)	12	12
Amphibious Command Ship (LCC)	24	24
Rotary-Wing Aircraft		
CH-46 Sea Knight	15	25
CH-53 Sea Stallion	24	*55
V22 Osprey	12	24
* With centerline seating installed		

Appendix G

Selection of Patients for Aeromedical Evacuation and Patient Classification Codes and Precedence

Patient classification codes provide information to evacuators and treatment personnel in an abbreviated form. They can be used to complete administrative reporting requirements pertaining to the evacuation of patients.

Paragraphs G-2 through G-5 implement STANAG 3204.

BRIEFING OF PATIENTS PRIOR TO AEROMEDICAL EVACUATION

G-1. When patients are evacuated by aircraft (routinely in Roles 2, 3, 4, and 5), they should be briefed on the following points—

- A number of ambulatory patients will be detailed to assist with the evacuation of litter patients in any emergency.
- Safety belts and litter straps are to be properly fastened according to orders given by the pilot.
- Patients are instructed on the proper position to assume in preparing for an emergency. Flight crews and CCATT ensure that seat safety harnesses have been tightened.
- Ambulatory patients, with the exception of those designated to assist litter patients, are the first to leave a downed aircraft.
- Immobilized litter patients are freed from litters and assisted in leaving the aircraft. Litters will not normally be removed from their fastenings in view of the limited time available to evacuate the aircraft.
- Mentally disturbed patients should be quieted so that the orderly removal of other patients will not be jeopardized.

INTERNATIONAL STANDARDIZATION EVACUATION AGREEMENT

G-2. Table G-1 provides the patient classification codes defined in international standardization agreements.

Table G-1. Patient classification codes

Code	*Classification*
Class 1. Psychiatric Category	
1A	Stretcher patients who are frankly disturbed and inaccessible, requiring close supervision and sedation, or the use of restraint equipment.
1B	Patients who are not at the time of request grossly disturbed, but who may react badly to air travel, or who may commit acts likely to endanger themselves or the safety of the aircraft and its occupants.
1C	Sitting patients who are cooperative and have proved themselves to be reliable under preflight observation.
Class 2. Stretcher Patients Other Than Psychiatric	
2A	Patients, who in an emergency, would be unable to leave the aircraft without assistance.
2B	Patients, who in an emergency, would be able to leave the aircraft unaided.
2C	Patients who can walk onto an aircraft but on a long flight would benefit from a litter.
Class 3. Sitting Patients Other Than Psychiatric	
3A	Sitting patients, including handicapped persons, who may need medical or nursing attention en route and who, in an emergency, would require assistance to escape.
3B	Sitting patients who may need medical or nursing attention en route and who would be able to escape unassisted in an emergency.
Class 4. Sitting Patients Other Then Psychiatric	
4	Patients who will not need medical or nursing attention en route and are capable of traveling unescorted. Class 4 patients are normally required to make their own way from the deplaning airfield to their destination.

INTERNATIONAL STANDARDIZATION EVACUATION PRECEDENCE

G-3. Patients for AE will be given appropriate degrees of priority so that, if aircraft space is limited, the more urgent patients may be evacuated before those whose conditions are less serious. The degrees of priority are depicted in Table G-2. (The evacuation precedence used by the USAF is essentially the same as this listing. It contains a few word changes and introduces specific time limits. It does not contain Priority 4.)

Table G-2. Patient priorities as designated in STANAG 3204

PRIORITY 1/URGENT	Emergency patients for whom speedy evacuation is necessary to save life or limb, to prevent complication of serious illness, or to avoid serious permanent disability.
PRIORITY 2/PRIORITY	Patients who require specialized treatment not available locally and who are liable to suffer unnecessary pain or disability unless evacuated with the least possible delay.
PRIORITY 3/ROUTINE	Patients whose immediate treatment requirements are available locally but whose prognosis would definitely benefit by air evacuation on routine scheduled flights.
PRIORITY 4	Patients for whom air evacuation is a matter of medical convenience rather than necessity.

PATIENT CLASSIFICATION

G-4. Table G-3 provides the patient classification codes used aboard USAF aircraft and that can be used in completing DD Form 601 (Patient Evacuation Manifest). These codes are expanded to include categories of patients and other personnel which may or may not apply on the battlefield (such as infants, relatives, or friends).

United States Air Force Evacuation Precedence

G-5. The evacuation precedence used by the USAF is dramatically different than that employed by the US Army medical evacuation system. These precedence should not be confused. Table G-4 (page G-5) provides the evacuation precedence and time frames used by the USAF.

Table G-3. Patient classification codes

Code	*Classification*
Class 1.	***Neuropsychiatric Patients***
1A	Severe psychiatric litter patients requiring the use of restraining apparatus, sedation, and close supervision at all times.
1B	Psychiatric litter patients of intermediate severity requiring tranquilizing medication or sedation, not normally requiring the use of restraining apparatus, but who react badly to air travel or who may commit acts likely to endanger themselves or the safety of the aircraft. Restraining apparatus should be available for use.
1C	Psychiatric walking patients of moderate severity who are cooperative and who have proved reliable under observation.
Class 2.	***Litter Patients (Other Than Psychiatric)***
2A	Immobile litter patients unable to move about of their own volition under any circumstances.
2B	Mobile litter patients able to move about of their own volition in an emergency.
Class 3.	***Walking Patients (Other Than Psychiatric)***
3A	Nonpsychiatric and nonsubstance abuse patients who require medical treatment, assistance, or observation en route.
3B	Recovered patients who are returning to their units and require no medical attention en route.
3C	Ambulatory drug or alcohol substance abuse patients.
Class 4.	***Infant Category***
4A	Infants under three years, occupying a seat or in a bassinet or a car seat secured in an ambulatory seat.
4B	Recovered infants under three years, occupying a seat or in a bassinet or car seat secured in an ambulatory seat.
4C	Infants in an incubator.
4D	Infants under three years on a litter.
4E	Outpatients under three years on a litter for comfort.
Class 5.	***Outpatient Category***
5A	Ambulatory, nonpsychiatric, and nonsubstance abuse outpatients who are traveling for an outpatient visit and do not require a litter or medical assistance in flight.
5B	Ambulatory drug or substance abuse outpatients going for treatment.
5C	Psychiatric outpatients going for treatment.
5D	Outpatients on a litter for comfort or safety.
5E	Returning outpatients on a litter for comfort or safety.
5F	Other returning outpatients.
Class 6.	***Attendant Category***
6A	Medical attendants, either physician, nurse, or technician, who are assigned to give specialized medical treatment or nursing care to a particular patient.
6B	Nonmedical attendants, either relatives or friends, who may assist with the patient's care and who may also require support.

Table G-4. Evacuation precedence used by the United States Air Force

Evacuation Precedence	*Definition And Time Frames*
URGENT PRECEDENCE	Applies only to the need for immediate life, limb or eyesight saving. These patients should be picked up and delivered to the designation facility as soon as possible.
PRIORITY PRECEDENCE	Applies to the need for prompt medical care not available locally. These patients should be picked up within 24 hours and delivered to the destination facility with the least possible delay.
ROUTINE PRECEDENCE	Applies to all other patients. Routine patients will be picked up and delivered on regularly scheduled flights.

Appendix H

Troop Leading Procedures, Precombat Checks/Precombat Inspections, and Leader Checklists

Troop leading procedures (TLPs) are universal in their application. They are used by leaders at all levels regardless of branch or MOS. For purposes of this appendix the leader titles of ambulance platoon leader and ambulance platoon sergeant are used for familiarity.

SECTION I — TROOP LEADING PROCEDURES

H-1. Leaders must conduct detailed planning and preparation before each mission in order to be successful. That is why the Army developed an eight-step process of TLP for leaders to follow.

H-2. Each step of the TLP process is designed to give leaders a framework on which to thoroughly develop their plans and orders. If these steps are not followed the resulting plans and orders may be missing information which may be critical to successfully execute the mission.

THE EIGHT STEPS OF TROOP LEADING PROCEDURES

STEP ONE: RECEIVE THE MISSION

H-3. Once staff planning is initiated the ambulance platoon leader gathers information concerning the operation from the company commander and disseminates to his subordinate leaders. Examples of this information include but are not limited to intelligence preparation of the battlefield (IPB) and METT-TC. If the company commander has not completed a full plan, he may still issue a warning order (WARNO) to the ambulance platoon leader.

- In keeping with the rules of time management, the ambulance platoon leader uses one-third of his overall time for planning the operation. The reverse planning sequence is recommended.
- The ambulance platoon leader's OPORD must also be understandable to subordinate personnel.
- Throughout the planning and development process the ambulance platoon leader may find it necessary to change or refine the plan as the tactical situation changes, as long as it follows the commander's intent.

STEP TWO: ISSUE A WARNING ORDER

H-4. The ambulance platoon leader issues a WARNO based on information received from the company commander's WARNO or OPORD.

- The ambulance platoon leader can present the order in any format.
- As a minimum, the platoon should use the company or battalion tactical standing operating procedures (TACSOP). The best method is to use already developed and implemented platoon level TACSOP that cover preparation for offensive or defensive missions. These TACSOPs should be thoroughly trained and rehearsed.
- Whatever the mission, the platoon can begin work on its preparation by assessing equipment needs through the various military supply classes (such as Class I, III, IV, V, or VIII).
- Every Soldier in the platoon should have a specific task and purpose (identified in the platoon SOP) once the ambulance platoon leader has issued a WARNO.

Step Three: Make a Tentative Plan

H-5. The ambulance platoon leader considers the following while preparing the tentative plan:

- Mission analysis, IPB, METT-TC considerations, higher mission and intent (two levels up).
- The ambulance platoon leader must know what systems are available for his use and how to employ them. He must fully understand his own equipment and capabilities as well as those of his enemy.
- Consider **Observation** and Fields of Fire, **Avenues** of Approach, **Key** Terrain, **Obstacles** and **Cover** and Concealment, (man-made and natural), (OAKOC), weather, morale, and flexibility to react to last-minute changes. As the commander updates the ambulance platoon leader, the ambulance platoon leader, in turn, continually updates the platoon with FRAGOs.
- Experience is a valuable asset for mental preparation. Knowing how Soldiers are going to react to situations, based on past experiences, can help any officer or NCO develop his plan.
- During the planning process, the leader should have a *vision* of the enemy and understand the threat using available intelligence. The better the ambulance platoon leader understands the enemy, the greater his opportunity for success.

Step Four: Start Necessary Movement

H-6. Even though listed as Step Four, necessary movement can begin as soon as the mission is received.

- If movement is required, the ambulance platoon sergeant, with the assistance of squad leaders, can move the platoon to a new tactical assembly area (TAA). This allows the ambulance platoon leader more time to prepare his order and graphics.
- At the new TAA, preparations continue with precombat checks (PCCs). This is a good opportunity to conduct PCCs on squad equipment by using the platoon SOP checklists and to certify squad rehearsals. If the ambulance platoon sergeant is unable to conduct squad inspections, the platoon SOP should designate someone else, such as the senior squad leader, for the task.
- If platoon linkups are time driven, movements and time management become critical. Failure to plan or allow for ample movement time creates late linkups and results in late rehearsals, late PCCs, and various other events that will not be accomplished on time or to standard.
- If movements were conducted late or linkups were late, the mission timeline will suffer.

Note. Time management is mission essential!

Step Five: Reconnoiter (Conduct Reconnaissance)

H-7. Although listed fifth, this step may be implemented at any time and as often as necessary during TLP. When the risk of enemy contact is too great, or establishing additional orders for reconnaissance parties is time consuming leaders may have to rely on map reconnaissance and or photographic imagery to reconnoiter an AO. Ambulance platoon leaders must ensure that ground evacuation routes are reconnoitered and strip maps developed for security escort vehicle drivers and ambulance drivers.

- A leader's reconnaissance must be performed! Leaders who see more of the AO are better able to develop more detailed and useful plans.
- Information gained by the leader's reconnaissance allows more accurate terrain models to be built and greater detail can be provided when the ambulance platoon leader is briefing the OPORD.

Step Six: Complete the Plan

H-8. Complete the plan—

- With reconnaissance completed, movement accomplished, and the initial plan established, the ambulance platoon leader should have most of the information that is needed to finish writing the OPORD.
- Any additional resources needed should be en route or already on-hand so the ambulance platoon leader, when briefing the OPORD, has positive control of additional equipment or manpower.
- Graphics must be finished before the OPORD is briefed to the platoon and quickly handed down to squad leader level. Logistics and sustainment graphics must include CBRN decontamination points, MTFs locations, and supported unit CCP and AXP.

STEP SEVEN: ISSUE THE ORDER

H-9. Avoid rushing through the OPORD briefing as this may result in pertinent information being left out of the brief.

- When completed, distribute the OPORD with graphics to each squad and team leader. Also include a terrain model or sand table to enhance visualization of the terrain.
- Minimize distractions. Having subordinate leaders in the right mindset to receive the order is critical to allow them to collect and retain all of the information presented at the OPORD brief.
- If possible, present the OPORD with a representation of the terrain in the background to enhance visualization of the mission.
- After the OPORD has been presented, the ambulance platoon leader ensures that the information has been disseminated to every Soldier. Backbriefs from Soldiers and NCOs to the ambulance platoon leader should reveal any portion or portions of the OPORD that were missed or not fully understood. Everyone leaving the OPORD brief should have a clear understanding of the mission.
- Contingencies need special consideration. If, during the OPORD brief, a potential activity is discussed that has no clear plan, subordinates can help develop a contingency plan, unless the activity is already specified in the platoon SOP. Defer making a decision on topics discussed during the OPORD brief which require more thought and possibly guidance from higher headquarters. If at all possible, allow a follow-up period to develop clarity on the situation and the techniques needed to accomplish the mission.

STEP EIGHT: REHEARSALS

H-10. Supervise rehearsals and plan to run as many rehearsals as time permits. After each rehearsal, conduct an after-action review (AAR). When conducting platoon rehearsals with attachments, ensure that the ambulance platoon leader/sergeant supervises the AAR with attachments. When conducting the AAR, ensure Soldier involvement. By this time all Soldiers understand the concept of the operation.

- When rehearsing the mission, ensure that the rehearsal lane coincides with what was briefed during the OPORD brief. Based on the latest intelligence, the rehearsal lane should match (as much as possible) what will be encountered during the actual mission. Do not allow for complacency. If there is information about the terrain and the obstacle that can be placed in the rehearsal lane to add realism, ensure that obstacle information is built into the lane. A company-level rehearsal is of limited value if the enemy obstacle does not resemble what was reconnoitered or templated.
- If time does not allow for a full rehearsal, backbriefs must be conducted. A successful understanding of the mission can be accomplished with key personnel by walking troops through the mission on the terrain model and asking questions. The backbrief is never a substitute for a rehearsal, but is an acceptable recourse when time is limited.
- Conduct final precombat inspections (PCI).
- If there are any last-minute concerns, they should be satisfied during this time.
- Conclude the brief.

H-11. Some steps of the TLPs require more development than others. Regardless of the mission, the ambulance platoon leader must not, for any reason, omit any step from the TLP.

SECTION II — PRECOMBAT CHECKS AND PRECOMBAT INSPECTIONS

H-12. Precombat checks and PCIs are tools which leaders have to ensure that their Soldiers—

- Fully understand the mission and their role.
- Have the equipment that is necessary to accomplish the mission on hand and that it is serviceable.

H-13. Leaders must conduct PCCs and PCIs for every mission. Failure to perform these vital checks could result in unnecessary casualties, loss of equipment, and failure to accomplish the mission.

H-14. When conducting PCCs/PCIs, as deficiencies are identified corrections must be made immediately.

H-15. Who performs PCCs and PCIs are both intended to accomplish the same thing, which is to have a leader ensure that personnel and equipment are ready. The only difference is who is doing the inspecting.

- Commanders conduct PCIs. Ambulance platoon leaders, ambulance platoon sergeants, and squad leaders conduct PCCs.
- Why do we identify a commander's inspection separately? Equipment readiness is essential for mission success. Squad leaders and ambulance platoon leaders are certainly capable of ensuring equipment readiness, but the commander's first-hand knowledge of mission requirements cannot be substituted.

H-16. Units need to establish standard PCC/PCI checklists based on their units mission profile. They should be included in the TACSOP and issued to all leaders. They should include routine daily checks and checks that are mission specific. They do not need to be extremely detailed; a list of items to check and the leader's knowledge of the equipment and the appropriate manuals should suffice. The PCCs checklists provided below are examples that units may use to develop their own or they can be modified as appropriate to meet their own requirements.

SECTION III — LEADER CHECKLISTS

H-17. This section provides sample checklists which may be used as is, or modified, to meet unit specific requirements. As with any checklist they are only effective if they are incorporated into a unit's TACSOPs and then used routinely.

GENERAL CONSIDERATIONS

H-18. Leader receives the mission from the next higher headquarters.

___ Clarifies any questions.

___ Coordinates with next higher headquarters as required.

___ Conducts mission analysis.

___ Produces the estimate.

___ Revises mission statement.

___ Produces tentative time schedule.

___ Issues a WARNO.

H-19. Unit members perform readiness, maintenance, and functional checks under the supervision of unit leaders which includes:

___ Medical equipment sets are present and all perishable and dated items are current and on-hand.

___ Vehicles/aircraft/generators.

___ Night vision devices.

___ Communications equipment.

___ Weapons and ammunition.

___ Field sanitation equipment and supplies.

___ Special equipment (such as hoist and forest penetrator).

___ Common table of allowances equipment.

H-20. Leader makes a tentative plan.

___ Uses estimate of the situation to depict plan of support.

___ Develops courses of action (COA).

___ War games COA.

___ Determines best COA.

___ Completes his plan.

H-21. Leader issues OPORD.

H-22. Leader uses sand table or sketches to depict plan of support.

H-23. Leader affects coordination for the mission.

___ Support requirements.

___ Current intelligence (to include health threat) update.

___ Control measures.

___ Communications and signal information.

___ Time schedule.

H-24. Leader receives attachments/augmentation (personnel), if appropriate.

___ Attachments are oriented to unit.

___ Attachments are briefed on the mission.

H-25. Leader supervises AHS mission preparation.

___ Key leaders brief back unit leader.

___ Key personnel rehearsals are conducted.

H-26. Unit leaders:

___ Supervise.

___ Inspect.

___ Ensure adequate security.

___ Conduct brief backs.

___ Rehearse.

___ Continue coordination.

H-27. Unit plans for support of combat operations.

___ Analyze mission.

___ Patient acquisition and medical evacuation requirements.

___ Area support requirements.

  ___ Requirements for water, Class VIII, and other supply classes.

  ___ Requirements for CBRN defense.

  ___ Transportation requirements.

H-28. Unit leaders:

  ___ Execute a work/rest plan based on work priorities and statutory (crew rest) requirements.

  ___ Monitor current situation.

  ___ Issues appropriate FRAGOs based on intelligence or operational updates.

  ___ React to messages or orders from higher headquarters.

  ___ Execute any actions and coordination resulting from change.

H-29. Unit headquarters remains current on positions and missions of higher, adjacent, and subordinate units.

Ground Ambulances Inspected For

___ Authorized MES are on hand.

___ Medical equipment is complete and serviceable.

___ Oxygen is on hand and serviceable.

___ Authorized medications are on hand and current.

___ Packing list is available.

___ Strip maps and/or road maps are available (with overlays).

___ On-vehicle equipment is on hand.

___ Log book is present and current.

___ All drivers are licensed.

___ Situational awareness equipment (position locator) is on hand and serviceable.

___ Communications equipment is on hand and serviceable and set to correct frequency.

___ Medical unit identification markers (according to the Geneva Conventions) are displayed.

Note. Markers are red on a white background only; camouflaged or subdued markers are not authorized.

Chemical, Biological, Radiological, and Nuclear Equipment

___ Individual protective equipment is on hand and serviceable. (One set is issued; the other is maintained in support.)

___ Protective masks are issued and serviceable.

___ Nerve Agent Antidote Kit is available and distributed, if required.

___ Convulsant antidote for nerve agent is available and distributed, if required.

___ Decontamination apparatus is complete and serviceable.

___ Basic load of decontamination supplies is on hand.

___ Chemical agent alarms are on hand and serviceable.

___ M256A1 detector kits are issued.

___ Chemical, biological, radiological, and nuclear contamination marking kits are distributed.

___ Chemical agent monitors are on hand, if authorized.

___ Replacement filters for protective masks are on hand.

___ Biological and chemical warfare agents prophylaxis, immunizations, and barrier creams have been accomplished, if appropriate.

___ Nerve agent pretreatment packets are available.

___ Radiac sets are on hand.

MISCELLANEOUS EQUIPMENT

___ Inspect binoculars.

___ Inspect camouflage nets and support systems, if appropriate.

___ Inspect night vision devices.

___ Ensure batteries are on hand and serviceable.

___ Inspect tentage.

___ Inspect global positioning systems, if available.

PERSONNEL

___ Ensure Soldiers are in the correct uniform.

___ Ask questions to ensure that Soldiers have been briefed on mission and situation.

___ Implement appropriate MOPP level.

___ Check for drivers licenses.

___ Brief Soldiers on operations safety and environmental injuries.

___ Individual equipment is on hand and stowed properly.

___ Soldier has eaten and is briefed on future field feeding.

___ Identification cards and tags are on hand and serviceable.

___ Camouflage self and equipment, when directed. (If medical vehicles are camouflaged, the Geneva Convention emblem must be removed.) (Refer to STANAG 2931.)

___ Work/rest plan (to include crew rest) is implemented.

___ Water discipline plan is implemented, if appropriate.

COMMUNICATIONS EQUIPMENT

___ Radios are operational (communications check completed).

___ Telemedicine equipment is available and operational, if available.

___ Speech security equipment functions, if available.

___ Frequencies are set.

___ Matching units are operational.

___ Antennas are tied down properly.

___ Connectors are clean and serviceable.

___ Field telephones (TA-312) are on hand and serviceable, if appropriate.

___ Batteries are on hand and charged.

___ Man-pack radio sets are complete.

___ Switchboard is on hand and serviceable.

___ Communications wire (WD-1) is on hand and serviceable.

___ Antennas and remotes are present and serviceable.

___ Signal operating instructions are available and secured.

___ Call signs, frequencies, and challenge passwords have been disseminated.

___ Perform communications check again.

VEHICLES

___ Loads are according to load plan; load plan is posted in the vehicle.

___ Hazardous cargo is properly identified and stored toward the rear of the vehicle for easy access and inspection.

___ Ammunition is issued and properly stored.

___ Vehicle fuel tank is topped off.

___ Package POL products and small arms lubricant are present.

___ Water cans are full.

___ Meals, read-to-eat are issued and stowed.

___ First-aid kits are present and complete.

___ Operator's manuals and lubrication orders are present for the vehicle, radios, and associated equipment.

___ Critical on-vehicle equipment and basic issue items are present.

___ Vehicle dispatch is complete.

___ Department of the Army (DA) Form 2404 (Equipment Inspection and Maintenance Worksheet) is complete.

___ No deadline deficiencies exist.

___ Before operation PMCS has been completed.

INDIVIDUAL WEAPONS

___ Clean and functional.

___ Cleaning tools/kits, bolts, and ruptured cartridge extractors are present.

___ Range cards are on hand.

___ Ammunition is issued, accounted for, and secured.

___ Magazines are issued.

___ Blank adapter installed, if appropriate.

___ Function check performed.

Medical Evacuation Support Plan

___ Mission statement/commander's intent/task organization.

___ Intelligence preparation of the battlefield considerations.

___ Patient estimates by phase.

___ Identify expected areas of patient density.

___ Anticipate and plan for likely patient conditions.

___ Formulate/evaluate/implement/reevaluate COA.

___ Location of supported units.

___ Location of supporting MTFs/FSTs/hospitals.

___ Location of supporting evacuation units.

___ Augmentation of personnel and vehicles coordinated supporting MTFs.

___ Class VIII supply/resupply coordinated and available.

___ Location of supported unit CCPs and AXPs.

___ Evacuation routes primary and alternate.

___ Trafficability (enemy activity, hazards, condition, obstacles).

___ Security (gunship escort, MP escort, convoy support).

___ Strip maps/overlays/maps.

___ Landing zones designated for rotary-wing evacuation aircraft (primary and alternate)

___ Ensure that platoon/company/battalion/treatment facility litter teams are available to assist ambulance crews in loading patients.

___ Available litters and types (NATO standard, folding nonrigid, SKED®, and so forth).

___ Ground ambulances available (type/number/carrying capacity).

___ Air ambulances available (type/number/carrying capacity).

___ Supporting evacuation assets available (requested/ prepositioned).

___ Nonmedical vehicles for CASEVAC (type/number/carrying capacity).

___ Available aircraft (UH-60, CH-47).

___ Medical evacuation support plan for units without evacuation resources.

___ Treatment and evacuation of EPW and detainees.

___ Chemical, biological, radiological, nuclear considerations (contaminated evacuation vehicle plan, patient decontamination site, decontamination sites, contaminated aircraft).

___ Other.

Casualty Care Considerations

___ Self-aid/buddy aid/CLS/trauma specialist.

___ Combat lifesavers trained and CLS bags stocked and issued.

___ Emergency care specialists and flight medics are rehearsed, aid bags fully stocked.

___ Trauma specialist Class VIII resupply sets fully stocked and pushed forward to company command post.

___ Locations of platoon/company/battalion/treatment facility CCPs are known.

___ Other.

Mass Casualty Plan (Resourced/Rehearsed/Coordinated)

___ Litter teams.

___ Litters (all).

___ Medical evacuation/CASEVAC vehicles (all).

___ Class VIII supplies (on hand).

___ Class VIII resupply (Prepackaged and prepositioned).

___ Company/battalion command net.

___ Brigade support medical company command net.

___ Personal effects, weapons, equipment, and property exchange.

___ Chemical, biological, radiological, and nuclear casualty/evacuation plan.

___ Location of decontamination site.

LANDING ZONE OPERATIONS

Safety

___ At all times take all the necessary safety precautions.

___ Platoon that is responsible for running the LZ provides all required safety gear. Included are—

 ___ Safety goggles.

 ___ Visual Signal, VS-17 panels/chemical lights.

 ___ Smoke (Appendix E).

___ Is the LZ secure.

Marking the LZ

___ Use colored smoke to identify the LZ and wind direction.

___ Use an inverted "Y" to mark the LZ.

___ During daylight, camouflage the LZ so the VS-17 panels cannot be seen until evacuation aircraft are confirmed and inbound.

___ At night, emplace chemical lights only when an AA is confirmed and is inbound.

Landing Requirements for Light Helicopters

___ Minimum real estate required is a cleared area 30 meters in diameter.

___ Area must have an approach and departure zone clear of obstructions and not over the established company area.

Appendix I

Safety on the Battlefield

The tactical environment provides ever-changing demands and unpredictable problems, often under stressful conditions. The interface of man, machine, and environment is constantly shifting. In this environment, mission accomplishment requires continuous leader involvement and flexible decision making. Not surprisingly, accidents and injuries increase during tactical operations. Safety in the tactical environment depends upon compliance with established standards. However, due to fluid conditions in the tactical environment, safe mission accomplishment relies heavily on complete integration of risk management into both planning and execution phases. Risk management assists commanders in anticipating and controlling hazards in the planning phase and in dealing with unexpected hazards as they arise in the execution phase.

SECTION I — GENERAL

COMMON ACCIDENTS

I-1. During all operations, it is critical commanders and leaders minimize risks through appropriate risk management procedures to protect the force and preserve the unit's war-fighting capability. Over half of all Army accidents during tactical operations are accredited to the following operations.

VEHICLE OPERATIONS

I-2. Most accidents in this category are caused by excessive speed for conditions (weather, traffic). Other causes are recklessness, fatigue, unfamiliarity with roads, and untrained and inexperienced drivers. A lack of equipment/vehicle handling characteristics knowledge also contributes to accidents. Strict enforcement of standards is needed for all vehicle operations. The senior occupant must be responsible for this enforcement. Only trained, licensed personnel should be assigned to operate vehicles or equipment. Ground guides are mandatory during movement in bivouac and assembly areas, when backing, and during periods of reduced visibility. Operators must be familiar with proper operation and maintenance of commercial equipment.

MATERIEL HANDLING

I-3. These accidents occur when an object is too large or too heavy to handle for those attempting to move it. As a result, someone may sustain a muscle or back injury, or may be crushed. Overconfidence in one's ability, a lack of planning, and fatigue are common factors in such accidents.

MAINTENANCE

I-4. These accidents are often caused by failure to follow procedures, using the wrong tools, and/or fatigue.

SEVERE ACCIDENTS

I-5. The following activities produce fewer accidents than those listed above; however, when they do occur, they often result in catastrophic damage or death. Commanders and leaders must pay special attention to unit missions which involve these activities or put Soldiers in situations or locations where they will come in contact with associated hazards and ensure they properly brief their personnel and mitigate risk to the fullest extent.

AMMUNITION AND EXPLOSIVES HANDLING

I-6. Horseplay, mishandling, disassembly, unauthorized use, and improper storage of ammunition and explosives account for many personnel injury accidents. It is essential to enforce accountability and security procedures for unexpended ammunition and explosives and to comply with explosive storage safety standards in AR 385–64 and DA Pam 385–64.

EXPLOSIVE SOUVENIRS

I-7. Educate Soldiers to dangers involved and the serious consequences of collecting unexploded ordnance on the battlefield or ranges. Post-tactical training shakedown inspections for this type of materiel are a must. Amnesty boxes are also useful. Platoon sergeants and squad leaders policing their Soldiers can prevent most of these accidents through proper training.

FIELD EXPEDIENTS

I-8. Be suspicious of shortcuts. Although tactical operations frequently involve employment of field expedients, risks and benefits must be carefully weighed. In many cases, field expedients are the result of a weak supply system or inadequate planning.

FIELD HEATERS AND STOVES

I-9. Operators of all types of heaters and stoves should be trained and licensed in advance. Equipment should be maintained and operated in accordance with operating instructions including use of proper fuel. Ensure combustible material is kept well away from heaters and stoves and fire fighting equipment is available for each heater and stove. Heaters or stoves with self-contained fuel supplies (small space heater) should not be refilled while the heater is on or still warm. Do not use heaters or stoves in tents or other confined spaces without use of proper ventilation, such as tent vent flaps, doors, or windows.

PETROLEUM, OILS, LUBRICANTS STORAGE AND HANDLING

I-10. The POL handlers must know and practice safety rules and procedures. Frequent inspections ensure safe storage and transfer of POL products. Proper grounding procedures must be followed. Field manual 10–67–1 describes use of protective equipment to prevent personnel exposure.

SOLDIER FATIGUE

I-11. When a Soldier's sleep time is dependent upon the tactical situation, debilitating fatigue can occur. Soldiers suffering from sleep loss experience various symptoms of fatigue including decreased coordination, narrowed attention span, and reduced performance to standard. Anticipate fatigue-related errors and take action to prevent them.

TACTICAL SLEEP PLAN

I-12. Commanders need to develop and enforce sleep plans. Control sleeping areas to prevent Soldiers from being crushed by moving vehicles in and out of the area.

Water Operations

I-13. Plan water operations carefully, as risks of drowning and equipment loss are high during water operations. Pair strong swimmers with weak ones to protect personnel. Secure equipment and float it across rather than requiring individuals to carry their equipment. Use safety lines and personal floatation devices.

Weapons

I-14. Most accidents occur when cleaning or clearing individual weapons, entering or exiting vehicles, or running with loaded weapons. Provide guidance for weapons handling, loading, and clearing, and ensure it is strictly enforced. Do not load weapons that are not essential for the current mission.

Weather-Related Casualties

I-15. Consider the effects of weather during planning. Unit effectiveness is quickly lost through weather-related casualties such as frostbite, heat stroke, lightning strikes, and falls. Instruct Soldiers in awareness, prevention, and first aid for weather-related injuries and when these conditions can be expected.

VEHICLE CONVOY OPERATIONS

I-16. Convoy operations can be very dangerous if not properly planned. Control of convoy speed and proper separation between vehicles is critical to reducing risk of an accident. See FM 4-01.45 for in-depth information on tactical convoy operations.

Convoy Commander Responsibilities

I-17. Convoy accidents are most commonly caused by a leader failing to perform his or her duties as the commander of a vehicle movement as required in FM 55-30. This includes failure to:

- Control the group's movement.
- Ensure vehicles maintain proper march speed.
- Properly mark unit vehicles.

I-18. Convoy commanders are responsible for ensuring the safe movement of the convoy. To do this, they must positively control the convoy's movement by using communications equipment among the vehicles. This includes ensuring each vehicle has a properly trained, equipped, and supervised crew leading from the front in the absence of radios and/or other means of managing the march. Control of the movement includes enforcing speed limits, march intervals, and seat belt usage. It also requires the ability to stop the march if an unexpected hazard is encountered along the route.

I-19. Preparing vehicles and Soldiers for movement is a leader responsibility. Inexperienced Soldiers, personnel turbulence, and ever increasing training requirements have caused some units to become complacent in managing risks associated with vehicle movements. According to the risk management process, leaders must identify hazards associated with the mission and develop, implement, and supervise control measures to mitigate those risks. These control measures include marking the vehicles in accordance with local SOPs, briefing crews on hazardous conditions expected along the route, precombat checks of personnel and equipment, and developing preaccident emergency contingency plans.

I-20. Unit SOPs should address leader responsibilities during movements of any number of vehicles and identify means to implement common controls.

Driver Skills

I-21. Operators should be taught the specific skills needed for tactical vehicle operations, which, in addition to requirements outlined in AR 600-55, include—

- Pulling and backing trailers.
- Vehicle recovery operations.

- Loading and lashing of cargo.
- Methods of negotiating difficult terrain such as sand dunes, rice paddies, mountainous terrain, icy roads.
- Ground-guide procedures and signals.
- Methods and procedures for retrieving vehicles stuck in snow, mud, sand, or other restrictive terrain.
- Proper parking and use of the proper-sized chock blocks.

Use of Safety Equipment

I-22. When the tactical situation allows, flashers should be turned on immediately if a vehicle is disabled or impedes traffic, and every effort made to move the disabled vehicle from the roadway. Highway warning kits should be provided for each vehicle in a convoy. When a vehicle is disabled, place the warning triangle a minimum of 100 meters to the rear of the vehicle and ensure personnel remain clear of the road and rear of the vehicle.

Night Operations

I-23. Personnel who are required to operate motor vehicles while wearing NVG must be trained and tested on use and operation of such devices, and the training recorded on the individual's driver training records. Other factors include:

- Ground guides should be used when moving vehicles at night in areas where troops are present.
- Blackout driving should be prohibited on roads open to the public.

FIRE PREVENTION

I-24. The risk of fire is high in areas where a large number of Soldiers are in tents. The following guidance will reduce the risk of fires:

- Establish a fire prevention and protection plan including procedures for inspecting and recharging fire extinguishers during tactical operations.
- Appoint a fire marshal for each bivouac area and train them in their duties. Train Soldiers in fire prevention techniques as well as emergency procedures in the event of a fire.
- Establish safe distances between tents to reduce risk of multiple losses from one fire.
- Provide available fire fighting equipment (portable extinguishers, sand, water buckets, and shovels) to contain small fires. Ensure personnel are trained on their use.
- Establish procedures for sounding fire alarms.
- Ensure no-smoking areas are established and enforced.
- Establish an inspection system ensuring compliance with fire prevention standards.
- Ensure flammable materials are stored according to appropriate directives and checklists.
- Ensure vehicle fire extinguishing/suppression systems are operational and crews are proficient in using the systems.
- Provide a designated fire plan, equipment, and trained personnel for POL storage, ammunition supply points, motor pools, hospitals, and hangars.

SECTION II — ASSEMBLY AREA OPERATIONS

SAFETY OFFICER/NONCOMMISSIONED OFFICER

I-25. The safety officer and NCO must establish, in coordination with the 1SG, platoon leaders, and platoon sergeants, procedures for ground guides and incorporation of these procedures into the unit SOP. Company commanders must weigh mission requirements and personnel available while in combat operations and use risk management analysis to validate procedures for implementation into the SOP.

Operating within the modular force structure may require assistance from support units to meet necessary personnel and vehicle requirements to safely ground guide and move aircraft.

GROUND VEHICLES

I-26. Commanders and leaders must be aware of hazards from ground vehicles in AAs. Vehicles moving in the AA should always have a ground guide leading the vehicle until out of the AA. This prevents injury to personnel or damage to equipment. Use of a one-way path through the unit area will assist greatly in preventing accidents. Additionally units may be operating in the vicinity of armored vehicles and Soldiers must take additional precautions due to the inability of armored vehicle crews to see personnel close to their vehicles.

I-27. During initial entry into the AA, platoon ground guides lead convoy parties to their platoon locations with each vehicle dismounting one person to ground guide their vehicle. During this period of movement into the AA, there is a higher probability of accident or injury due to the amount of traffic and unfamiliarity of the area. Commanders must ensure movement into the AA is thoroughly briefed.

AIRCRAFT

I-28. Aircraft in the AA present movement hazards and contribute to an additional hazard of blowing debris. Blowing debris is a hazard to personnel, aircraft, and equipment. The variation in environments in which a unit may operate determines the type and amount of debris. Desert environments, for instance, produce brownout conditions which are a hazard to personnel, equipment, aircraft, and crews. Preventive measures should be incorporated into the unit SOP mitigating this risk. An example of risk mitigation for aircraft operations, in and around the AA parking areas, is brownout mats. These mats reduce the brownout effect of operating in a desert environment. Commanders must ensure SOPs are implemented for each of the environments the unit may operate within. This will allow the unit to establish ground guide procedures for aircraft, whether while hovering into and out of the AA or during ground movement of the aircraft within the AA.

I-29. Ground guide procedures for flight into or out of the AA will be based on the distance between aircraft and proximity to hazards (natural or manmade). In environments where the AA allows for wide scatter of aircraft, the need for ground guides while hovering into and out of the parking area is negligible. In contrast, when the AA parking areas require close interval spacing of aircraft, the need for guides is critical to safety. Additionally, if the AA requires aircraft to park in close proximity to natural or manmade hazards, ground guides are required. The need for ground guides is more critical with operations under night or with NVG devices.

I-30. Upon initial entry into the AA, ground guides should meet incoming aircraft. The situation may dictate crews entering the AA know limited information on where they will park their aircraft. The ground guides guide the aircraft into their assigned parking spot. In desert environments, the ability of personnel to guide aircraft into the parking location is decreased. Brownout upon landing requires a thorough prebrief to the flight crew on the AA parking plan and as exact a location as possible in relation to a ground marking, such as an inverted Y, of each aircraft parking location. Certain situations may require aircraft to land directly on a parking spot. The company safety officer and NCO must be involved in the design and procedures planned for aircraft operations into and out of the AA.

WING WALKING

I-31. Ground guides will also be utilized while moving aircraft with ground handling equipment. Ground guiding aircraft while moving with a vehicle requires additional personnel to ensure rotor blades and aircraft do not come in contact with another aircraft or object. Commanders must ensure procedures designed by the safety officer/NCO are integrated into the SOP. The plan should address guides necessary for various settings (such as define how many are necessary for close proximity parking and scattered parking not in close proximity to another aircraft or hazard). Ground movement of aircraft is normally associated with aviation maintenance operations. The maintenance officer should be involved in the design of ground guide/wing walking procedures for the unit SOP. A critical component of the aircraft ground movement plan is communication. Personnel conducting these operations must conduct a team prebrief and ensure that the team leader has a device that will signal all members to immediately halt the operation (survival vest whistle).

GROUND GUIDES

I-32. Commanders and leaders must ensure ground guide operations, vehicle and aircraft, are incorporated into their company SOPs. During AA operations, ground guides perform a key role in preventing personnel injury, and vehicle and equipment damage. Ground guide operations are critical to maintaining combat readiness for the unit.

PYROTECHNICS

I-33. Aviation units normally use pyrotechnics while in an AA. Units utilize pyrotechnics along likely enemy avenues of advance within their security sector as a warning of activity along these avenues. Additionally, aircrews utilize pyrotechnics as signaling devices in downed aircrew rescue operations. Commanders, unit leaders, and the ALSE officer must train personnel in proper use of pyrotechnics and its associated hazards. These items will be addressed in the unit SOP.

I-34. Pyrotechnics require protection against moisture, dampness, and high temperature. Pyrotechnic items must be given high priority for the best available protection because of their sensitivity. Wet pyrotechnic material is hazardous to store; consequently, any boxes showing signs of dampness will be removed from a storage site and inspected. If the pyrotechnic materiel is wet, it will be destroyed. Certain kinds of materiel deteriorate with age and have expiration dates on the containers. Loose pyrotechnic tracer composition, flare composition, and similar mixtures spilled from broken containers should be carefully taken up and covered completely with SAE 10 (EO–10) engine oil and removed for appropriate disposal.

I-35. Pyrotechnics are hazard division 1.3 which includes items that burn vigorously and cannot usually be extinguished in storage situations. Explosions are confined to pressure ruptures of containers and do not produce propagating shock waves or damaging blast overpressure beyond the magazine distance specified in DA Pam 385-64, Table 5–16. Tossing about of burning container materials, propellant, or other flaming debris may cause a severe hazard of spreading fire.

I-36. The fire fighting extinguishing agent for pyrotechnic materials, Class D (combustible metals [magnesium, potassium, and so forth]) hazard, should be a dry powder agent. Pyrotechnics are a mass fire hazard due to their chemical agents. The mass fire hazard can be more hazardous based on other likely types of ammunition stored with pyrotechnics. These fires may be fought if:

I-37. Explosives are not directly involved.

- White phosphorous munitions are involved, there may be an explosion (immerse/spray with water continuously).
- Pyrotechnics and magnesium incendiaries do not use carbon dioxide, Halon extinguishers, or water on or near munitions. Allow magnesium to cool unless upon flammable materiel. In this case, use a 2-inch layer of dry sand or powder on the floor and rake the burning material onto this layer and re-smother.

WARNING

DO NOT DISCHARGE PYROTECHNICS IN THE VICINITY OF AIRCRAFT FLYING IN THE AREA.

SECTION III — PROTECTING THE SOLDIER

ENVIRONMENTAL CONSIDERATIONS

I-38. As the military looks into the next century, and even today, military units should try to avoid unnecessary environmental damage, not only in training, but also across the spectrum of operational missions. A mission's success may be determined by political or socio-economic stability, both of which are affected by environmental factors and resources. The Army must be able to identify ways to protect the natural environment while executing full range of their missions:

- Considering the environment in planning and decision making in conjunction with other essential considerations of national policy.
- Protecting the environment of home stations and training areas as a means of retaining resources for mission purposes.
- Using environmental risk assessment and management principles to integrate environmental considerations into mission performance.
- Instilling an environmental ethic in Soldiers.
- Understanding linkages between environmental protection issues and their associated impact on safety, force protection, and AHS.

I-39. Soldiers must be prepared to respond across full spectrum operations in any component (offense, defense, stability or civil support operations), sometimes within the same operation. Deployed forces must be able to conform to environmental protection requirements of the theater commander without impairing combat effectiveness. For in-depth information on environmental considerations and Soldier protection, see FM 3-100.4.

I-40. When deployed, company commanders will often deal with base camps. Base camps, while not installations, are comparable to small towns and require many of the considerations applied to installations. A mayor (often the HHC commander) assists base camp commander with control of base operations. A base camp coordination agency will provide expertise and support to the commander, largely through its subordinate base camp assistance/assessment team. Environmental expertise is resident or aligned with this team and available to support the base camp commander and designated mayor of the base camp. They provide technical recommendations and maintain appropriate standards. More information about this phenomenon is provided in the Center for Army Lessons Learned Newsletter 99-9, Integrating Military Environmental Protection.

I-41. Army Regulation 200-1 and the Army's Commander's Guide to Environmental Management specify commanders' environmental responsibilities. To carry out these responsibilities, commanders:

- Comply with an installation's environmental policies and legally applicable and appropriate federal, state, and local laws and regulations or country-specific final governing standards if outside the continental United States.
- Demonstrate a positive and proactive commitment to environmental stewardship and protection.
- Provide environmental training required by law, regulation, or command policy.
- Ensure all personnel perform their duties in compliance with environmental laws and regulations, and respond properly to emergencies.
- Promote proactive environmental measures and pollution prevention.
- Supervise compliance with environmental laws and regulations during operational, training, and administrative activities.
- Include environmental considerations in mission planning, briefings, meetings, execution, and AARs.
- Understand the requirements of Army environmental programs.
- Identify and assess environmental risks of proposed programs and activities.
- Coordinate unit activities with higher headquarters environmental elements.
- Appoint and train an environmental compliance officer (ECO) coordinator for the unit.

- Ensure SOPs contain all environmental considerations and regulatory requirements appropriate for the level of command.
- Conduct environmental self-assessment or internal environmental compliance assessments.
- Understand linkages between environmental considerations and their associated impact on safety, force protection, and AHS.

ENVIRONMENTAL DUTIES

I-42. The key to fulfilling environmental requirements successfully at unit level is the ECO. Army Regulation 200-1 directs all Army unit commanders "appoint and train ECOs at appropriate levels to ensure compliance actions take place." In company-sized units, this duty will generally translate into an extra duty. The ECO manages environmental issues at unit level and ensures environmental compliance. He also coordinates, through the respective chain of command, with the supporting installation environmental staff to clarify requirements and obtain assistance.

I-43. The ECO accomplishes environmental compliance requirements on behalf of the commander. He also coordinates with supporting installation environmental staff to clarify requirements and obtain assistance. While this position of responsibility is not a formal staff position, the ECO is critical to the commander's environmental program. The ECO:

- Advises the unit on environmental compliance during training, operations, and logistics functions.
- Serves as commander's eyes and ears for environmental matters.
- Coordinates between unit and higher/installation headquarters environmental staffs.
- Manages information concerning unit's environmental training and certification requirements.
- Performs unit environmental self-assessment inspections.
- Performs environmental risk assessments.

UNIT PLANNING

I-44. Staffs integrate environmental protection into planning for larger units. Unit leaders integrate environmental protection into unit planning for battalion- and company-level units. Soldier safety should be addressed in SOPs, OPORDS, and training plans.

STANDING OPERATING PROCEDURES

I-45. Unit leaders develop SOPs reflecting environmental protection considerations for routine tasks and activities. The SOPs provide information to Soldiers on how to accomplish routine tasks in an environmentally sound manner. The SOPs incorporate local requirements. As local requirements change, unit leaders update their SOPs. The SOPs also help define environmental protection requirements for all unit activities—facility operations, field operations, deployment, and combat. Unit leaders ensure SOPs comply with local requirements by coordinating with higher headquarters staff—usually the environmental office, the surgeon and his staff, preventive medicine personnel, and the SJA or engineer coordinator.

I-46. Unit leaders conduct environmental risk assessments when planning operations or activities. Risk assessment is a standard element of TLP. Unit leaders perform environmental risk assessments for activities not addressed in the SOP or when conditions differ significantly from those described in the SOP. A maintenance unit does not perform a risk assessment every time it performs a lubrication or service. Rather, the SOP describes the correct manner to perform these actions. Risk assessments apply to garrison operations as well as field operations.

OPERATION ORDERS

I-47. Unit leaders address environmental protection in their plans and orders including: WARNOs, OPORDs, OPLANs, concept of operation plans (CONPLANs), and FRAGOs. The higher headquarters staff develops an environmental appendix/annex to its OPORD/OPLAN/CONPLAN. Subordinate unit leaders draw environmental information from the environmental appendix to the OPORD/OPLAN/

CONPLAN, or from Annex L in a joint operations planning and execution system document. Field Manual 5-0 directs the inclusion of Appendix 2 (Environmental Considerations) to Annex F (Engineer) of the OPLAN/OPORD/CONPLAN (battalion level and above) and specifies that lower-level unit leaders/staffs include environmental information in coordinating instructions and service and support paragraphs.

Risk Management Plan

I-48. Risk management is an integral part of the military decision-making process. The risk management process provides the framework for making risk management a routine part of planning, preparing, and executing operational missions and everyday tasks. Assessing environmentally-related risks is part of the total risk management process.

I-49. Knowledge of environmental factors is a key to planning and decision making. With this knowledge, leaders quantify risks, detect problem areas, reduce risk of injury or death, reduce property damage, and ensure compliance with environmental laws and regulations. Unit leaders should conduct risk assessments before conducting any training, operations, or logistical activities.

Tactical Risk/Accident Risk

I-50. When assessing the risk of hazards in operations, the commander and leaders must look at two types of risk.

- Tactical risk is risk concerned with hazards existing because of the presence of either the enemy or an adversary, thus involving the considerations of force protection. It applies to all levels of war and across the spectrum of operations. For example, during the Gulf War, the enemy's demolition of oil fields created a significant health and environmental hazard to the surrounding countryside and those units maneuvering through the area.
- Accident risk includes all operational risk considerations other than tactical risk. It includes risk to friendly forces and risk posed to civilians by an operation, as well as the impact of operations on the environment. It can include activities associated with hazards concerning friendly personnel, civilians, equipment readiness, and environmental conditions. Examples of environmentally related accident risk are improper disposal of hazardous waste, personnel not properly trained to clean up a spill, and units maneuvering in ecologically sensitive terrain. Preventive medicine considerations also fall into this area of risk.

Note: Tactical risk and accident risk may be diametrically opposed. The commander may accept a high level of environmentally-related accident risk on reducing the overall tactical risk. For example, a commander may decide to destroy an enemy's petroleum storage area to reduce its overall tactical risk.

Training Plans

I-51. Formal planning for training culminates with publication of the training schedule. Informal planning and coordination (preexecution checks) continue until training is performed. During rehearsals, leaders ensure all safety and environmental considerations are met.

I-52. To conduct effective, meaningful training for Soldiers, leaders, and units, thorough preparation is essential. This means well prepared trainers, Soldiers, and support personnel are ready to participate, and their facilities, equipment, and materials are ready to use.

I-53. A unit executes training the same way it executes a combat mission. The chain of command is present, in charge, and responsible. During operations, leaders ensure environmental practices and preventive measures are being employed. Once Soldiers understand what is expected of them, these practices become merely another measure of unit proficiency and level of unit discipline.

FIGHTER MANAGEMENT

I-54. Fighter management describes a process essential to effective and efficient battle staff operations. At company level, commanders do not have staffs. The battle staff becomes those personnel in the unit who operate the CP. Successful continuous operations at the company level are more demanding than higher level organizations. The unit requires a TACSOP allowing rest, especially for critical personnel. At company level, the commander and 1SG will find it very demanding to try to rest. Their rest plan must be a priority for the organization to be lead effectively and their units to be successful on the battlefield.

I-55. The cycle of recurring events within a CP focuses the CP officer in charge/noncommissioned officer in charge (OIC/NCOIC) on meeting information and action requirements. Company personnel are normally required to attend battalion/company recurring meetings. This has an impact on company CP operations and schedules. These recurring events include—

- Shift changes.
- Battle update briefings to the commander.
- Battalion targeting meetings.
- Battalion reports.
- Battalion battle updates without the commander.
- Battalion commanders' collaborative sessions.
- Company CP OIC/NCOIC shift change briefing/collaborative sessions.

I-56. The company CP OIC/NCOIC must achieve battle rhythm for updating and viewing information and understand how to use it to affect operations. A well-established battle rhythm aids the commander and leaders with CP organization, information management and display, decision making, and fighting the battle. Battle rhythm demands careful planning and design. Many competing demands must be deconflicted. Even subordinate platoons and sections affect a company battle rhythm based on their needs and unit procedures.

I-57. Command discipline is required to enforce sleep cycles and create an environment where sound sleep can be achieved. Battle rhythm and associated planning is a proactive, not reactive, exercise.

SLEEP PLANS

I-58. Units must develop detailed rest plans and enforce them. Leaders must rest to maintain their effectiveness; however, some leaders attempt to get involved in every aspect of planning and execution. An integral part of the planning process is determining when senior leader presence is required. It is just as important to identify when a leader's presence is not required. The planning process should include the following supporting techniques:

- Include a sleep plan in the METT-TC analysis.
- Ensure leaders have confidence in second and third echelons of leadership and their ability to make routine decisions.
- Instill trust and confidence in the officers, junior NCOs, and specialists by effective training and SOPs.
- Consider contingencies and establish criteria for waking leaders.
- Post sleep plans in CPs, platoon, and section areas; everyone needs to know the plan.
- Synchronize sleep plans with higher headquarters and subordinate platoons/sections.

CONTINUOUS OPERATIONS

I-59. Continuous operations are combat operations continuing around the clock at a high pace, requiring Soldiers to fight without relief for extended periods. Opportunities for sleep are scattered throughout the day and night.

SUSTAINED OPERATIONS

I-60. Sustained operations are operations conducted 24 hours a day with little or no opportunity for sleep. Sustained operations are when the same Soldiers or small units engage in combat operations with no opportunity for the unit to stand down and little time for Soldiers to sleep. Aviation units routinely conduct sustained operations to an extent even greater than the infantry due to aviation's maneuver, maneuver support and sustainment roles. To combat the negative effects of sustained operations, commanders and leaders must enforce fighter management principles on their organizations. A fighter management policy should be addressed in the unit SOP, and higher headquarters policy should correlate with company policy.

DEGRADATION OF COMBAT CAPABILITY

I-61. As sustained operations continue, all Soldiers begin to show effects of general fatigue and lack of sleep. Unless counteracted, unit performance of combat tasks decline. Recent studies indicate performance is degraded by 25 percent for each 24-hour period without sleep. After 96 hours, performance can be expected to be near zero. Determination to endure must be supplemented by countering the adverse effects to slow the rate of decline. It becomes more difficult to perform assigned tasks to the required standard. Leaders need to recognize signs of serious sleep deprivation in their subordinates. Studies show performance in all duty positions does not degrade the same. Performance in a duty position where there is a heavy load of mental tasks, such as aviation duties (determining, calculating, thinking, decision-making), degrades faster than the performance in a position whose tasks are mainly physical (firing, running, lifting, digging).

I-62. In addition to the degradation caused by fear, fatigue, and loss of sleep, there is a significant loss of effectiveness caused by operation in MOPP 4. When units are using full CBRN protective equipment, judgment is degraded, communications are less effective, and information flow between units is reduced.

FATIGUE IN FLYING OPERATIONS

I-63. Aviation operations are inherently dangerous. Commanders and leaders must be aware of increased dangers present in combat aviation continuous and sustained operations. Additionally, leaders must be able to recognize the symptoms of fatigue and how to deal with fatigued aircrews. Field Manual 3-04.301 provides an in-depth review of aeromedical factors associated with aviation.

OPERATIONAL TEMPO AND BATTLE RHYTHM

I-64. The aviation company should be staffed for 24-hour operations; however, it also conducts cyclical missions. Standard operating procedures establish methods of ensuring the right personnel are available for either cyclical or 24-hour operations. Regardless of methods used, practice during exercises must determine strengths and weaknesses of headquarters personnel for CP operations and training for additional personnel who may be utilized to staff the CP during continuous or sustained operations. Such knowledge allows leaders to focus on critical areas and personnel requiring additional training.

I-65. In planning, the battalion staff must consider battle rhythm requirements of subordinate companies and platoons. Depending on the situation, the battalion may schedule missions allowing company or platoon rotations to maintain their battle rhythm.

ABSENCE OF BATTLE RHYTHM

I-66. Without procedures establishing battle rhythm, leaders and units reach a point of diminished returns. This typically occurs between 72 and 96 hours of operations. As leader fatigue sets in, information flow, planning process, execution, and sustainment suffer—often greatly. Symptoms of diminished battle rhythm include the following:

- Leader fatigue.
- Leaders not fully aware of critical decision points (DPs).
- Leaders not available at critical DPs.
- Disjointed time lines between various levels of command.

Presence of Battle Rhythm

I-67. Battle rhythm allows units and leaders to function at a sustained level of efficiency for extended periods. Effective battle rhythm permits an acceptable level of leadership at all times. It can focus leadership at critical points in the fight or during particular events. Procedures and processes facilitating efficient decision making and parallel planning are critical to achieving battle rhythm. Every component of battle rhythm contributes uniquely to sustained operations.

Training

I-68. It is difficult, if not impossible, to establish battle rhythm while simultaneously conducting operations. Preplanning makes it happen. Planning, preparing, and training before deployment lays a solid foundation for viable battle rhythm during operations. Commanders must ensure all company personnel are trained on TOC set-up and operations, and they understand the team concept required to conduct CP operations during combat operations for extended periods. Additionally, company commanders must coordinate with battalion staffs during training to deconflict staff operations and requirements which are counterproductive to the company battle rhythm.

Battle Rhythm Elements

I-69. Battle rhythm is a multifaceted concept including the following elements:

- Sleep/rest plans.
- Trained second- and third-tier leadership in CPs.
- Synchronized upward multiechelon timelines.
- Parallel planning.
- Established processes and SOPs.

Command Post Personnel Depth

I-70. Established processes and SOPs relieve many antagonistic effects of extended operations. Standard operation procedures establishing and maintaining battle rhythm by facilitating routine decisions and operations are a step in the right direction. Soldiers trained to act appropriately in the absence of leaders or orders can relieve commanders and leaders of many of the time-consuming tasks robbing them of essential rest. Examples of areas NCOs and junior officers can accomplish for the commander include—

- Battle summaries and updates during a fight.
- Intelligence updates before, during, and after a battle.
- Sustainment updates before, during, and after a battle.
- Updates to the next higher commander.
- Shift change briefings.

Challenges to Battle Rhythm

I-71. Challenges to battle rhythm include enlisted, NCO, and junior officer duties; and synchronization of planning, execution, rehearsal timelines, equipment maintenance, and sleep plans.

Noncommissioned Officer and Junior Officer Responsibilities

I-72. At the company level, all personnel are critical to the success of the operation and provide valuable contributions. It is imperative commanders make certain each Soldier understands the importance of even the most menial tasks, such as CP security and TOC setup and teardown. Given the amount of personnel in a company, 24-hour operations and force protection requirements, commanders must utilize all personnel available to successfully accomplish CP operations and continuous missions. This becomes even more critical when a company has platoons or sections staged forward in support of BCTs. The improper use of personnel produces the following results:

- Key leaders become exhausted.

- CP operations trained personnel become exhausted.
- The initiative of trained subordinates is stifled, and the incentive to train is diminished.

I-73. The following techniques ensure proper use of personnel:

- Appropriate tasks are assigned to junior NCOs and specialists.
- Effective training and SOPs instill trust in officers and confidence in junior NCOs and specialists.
- Effective command guidance to the company that it is a team effort and all personnel available contribute to the effort.

CONTINUOUS OPERATIONS AND TIMELINES SYNCHRONIZATION

I-74. Timelines for the operation at hand must allow for not only the next operation, but also extended continuous operations. Synchronized, multiechelon timelines assist units in achieving battle rhythm. If units do not address critical events at least one level up and down, disruption results. An example of an unsynchronized timeline is a battalion rehearsal that conflicts with platoon precombat inspections or other events in their internal timeline. Lower echelon units seldom recover from a poor timeline directed by a higher headquarters, as well as platoons and sections at company level. Company commanders must coordinate with their battalion staffs on development of SOPs that include planning, rehearsal, and execution timelines one level below battalion to prevent these conflicts.

STANDING OPERATING PROCEDURES UTILIZATION

I-75. SOPs must be practiced and reviewed during professional development and sergeants' time. The existence of an SOP will not resolve troop-leading challenges unless the SOP is practiced often and internalized by unit members. Checklists are critical, as many leaders will often find themselves rushed, physically fatigued, distracted, and deprived of sleep. Checklists ensure each step is considered even when leaders are exhausted.

SECTION IV — ACCIDENT/INCIDENT REPORTING

I-76. Persons involved in, or aware of, an accident must immediately report it to the commander or supervisor directly responsible for the operation, materiel, or person(s) involved. Table I-1 and Table I-2 identify reporting requirements for aviation and ground accidents/incidents (peacetime and combat). See AR 385-40, DA Pam 385-40, and AR 385-95 for in-depth instructions on Army accidents. Company commanders ensure unit SOPs cover accident/incident reporting, and it is fully addressed by the company aviation safety officer.

Table I-1. Aviation accident notification and reporting requirements and suspenses

Peacetime				***Combat ****	
Accident Class	***Notification Telephonic Worksheet***	***DA Form 2397***	***Reporting AAA report***	***Notification Telephonic Worksheet***	***Reporting AAA report***
A	Immediate–to USASC (telephonic notification–no hardcopy notification required (DSN 558-2660/2539/3410 or Commercial (334) 255-2660/2539/3410	(CAI/IAI) 90 calendar days	Aircraft ground ACDTS only-30 calendar days	Same as peacetime to USASC or Safety Rep, forward	(Only when CDR determines DA Form 2397 investigation/ report not feasible) submit as soon as conditions, situation permit- Do not exceed 30 calendar days.
B	Immediate–to	(CAI/IAI) 90	Aircraft ground	Same as	(Only when CDR

Peacetime				***Combat ****	
Accident Class	***Notification Telephonic Worksheet***	***DA Form 2397***	***Reporting AAA report***	***Notification Telephonic Worksheet***	***Reporting AAA report***
	USASC (telephonic notification–no hardcopy notification required (DSN 558-2660/2539/3410 or Commercial (334) 255-2660/2539/3410	calendar days	ACDTS only 30 calendar days	peacetime to USASC or Safety Rep, forward	determines DA Form 2397 investigation/ report not feasible) submit as soon as conditions, situation permit- Do not exceed 30 calendar days.
C	Immediate–to USASC (telephonic notification–no hardcopy notification required (DSN 558-2660/2539/ 3410 or Commercial (334) 255-2660/2539/3410	N/A	30 calendar days	Same as peacetime to USASC or Safety Rep, forward	Same as peacetime
D	N/A (Unless SOF issue involved/ suspected)	N/A	10 calendar days	Same as peacetime	Same as peacetime
E	N/A (Unless SOF issue involved/ suspected)	N/A	10 calendar days	Same as peacetime	Same as peacetime
F	N/A (Unless SOF issue involved/ suspected)	N/A	10 calendar days	Same as peacetime	Same as peacetime
Submis-sion Methods	Class A-C telephonic (immediate) Class D, E, F-IF SOF	Mail	Typed or hand printed AAA reports by mail/FAX or courier/ message/ format/elec-tronic submission, Include attachments as required	Same as peacetime	Same as peacetime
Notes * Only when the senior tactical commander determines that the situation, conditions, and/or time does not permit normal peacetime investigating and reporting.					

Table I-2. Ground accidents notification and reporting requirements and suspenses (3)

Peacetime				*Combat (2)*	
Accident	***Telephonic Notification***	***AGAR***	***DA Form 285***	***Telephonic Notification***	***AGAR only*** ***By any means possible***
Class	***Worksheet***			***Worksheet***	***(Message, Electronic, FAX, Phone, Hand Carry, Mail)***
On-Duty					
A	Immediately (1)	Not required	IAI/CAI-90 days	Immediately (1)	As time permits (Not to exceed 30 days)
B	Immediately (1)	Not required	IAI/CAI-90 days	Immediately (1)	As time permits (Not to exceed 30 days)
C	Not required	Within 30 days	Not required	Not required	As time permits (Not to exceed 30 days)
D	Not required	Within 30 days	Not required	Not required	As time permits (Not to exceed 30 days)
Off-Duty					
A	Immediately (1)	Within 30 days	Not required	Immediately (1)	As time permits (Not to exceed 30 days)
B	Immediately (1)	Within 30 days	Not required	Immediately (1)	As time permits (Not to exceed 30 days)
C	Not required	Within 30 days	Not required	Not required	As time permits (Not to exceed 30 days)
D	Not required	Within 30 days	Not required	Not required	As time permits (Not to exceed 30 days)

Notes:

(1) USASC must be notified IMMEDIATELY by phone at DSN 558-2660/2539-3410 or Commercial (334) 255-2660/2539/3410 or notify USASC Safety Rep forward (during combat).

(2) ONLY when the senior tactical commander determines that the situation, conditions, and/or time does not permit normal peacetime investigating and reporting.

(3) Army civilian injury only accidents should be reported on appropriate Department of Labor Forms according to this regulation.

Glossary

SECTION I — ACRONYMS AND ABBREVIATIONS

A2C2	Army airspace command and control
AA	air ambulance/active Army
AAA	abbreviated aviation accidents
AAR	after-action review
AATF	air assault task force
ABCA	American, British, Canadian, and Australian
ACDTS	accidents
ACR	armor cavalry regiment
AD	air defense
AE	aeromedical evacuation
AECT	aeromedical evacuation control team
AELT	aeromedical evacuation liaison team
AF	Air Force
AGAR	abbreviated ground accident report
AGPU	aviation ground power units
AHS	Army Health System
ALCT	airlift control team
ALSE	aviation life support equipment
amb	ambulance
AMEDD	Army Medical Department
AO	area of operations
AOC	air operations center
AOE	Army of Excellence
APOD	aerial port(s) of debarkation
AR	Army regulation
ASI	additional skill identifier
ARSOF	Army special operations forces
ARNG	Army National Guard
ASCC	Army service component command
ASF	aeromedical staging facility
ASMC	area support medical company
ASOS	Army support to other Services
ASTS	aeromedical staging squadron
ATM	advanced trauma management
ATS	air traffic service

attn	attention
aug	augmentation
avail	available
AVIM	aviation intermediate maintenance
AVN	aviation
AVUM	aviation unit maintenance
AXP	ambulance exchange point
BAE	brigade aviation element
BAS	battalion aid station
BCT	brigade combat team
bde	brigade
BH	behavioral health
BIFV	Bradley infantry fighting vehicle
bn	battalion
BOA	basis of allocation
BSA	brigade support area
BSB	brigade support battalion
BSMC	brigade support medical company
BSS	brigade surgeon section
BTU	British thermal unit
C2	command and control
CA	civil affairs
CAB	combat aviation brigade
CAI	centralized accident investigation
CASEVAC	casualty evacuation
CASF	contingency aeromedical staging facility
CBRN	chemical, biological, radiological, and nuclear
CBRNE	chemical, biological, radiological, nuclear, and high yield explosives
CCATT	critical care air transport team
CCIR	commander's critical information requirements
CCP	casualty collecting point
CE	communications-electronic
CJA	Command Judge Advocate
CJTF	combined joint task force
CLS	combat lifesaver
CLT	casualty liaision team
CM	campaign module
CMCC	corps movement control center
CMD	command
CMO	civil-military operations

CO	company
COA	course(s) of action
COCOM	combatant command
COMSEC	communications security
CONPLAN	concept of operation plan
CONUS	continental United States
COSC	combat and operational stress control
COSR	combat and operational stress reaction
CP	command post
CRTS	casualty receiving and treatment ships
CSH	combat support hospital
CSAR	combat search and rescue
CTC	Combat Training Center
CZ	combat zone
DA	Department of the Army
DCO	defense coordination officer
DCSCOMPT	Deputy Chief of Staff, Comptroller, US Army
DCSIM	Deputy Chief of Staff for Information Managment, US Army
DCSLOG	Deputy Chief of Staff for Logistics, US Army
DCSPER	Deputy Chief of Staff for Personnel, US Army
DCSSPO	Deputy Chief of Staff for Support Operation, US Army
DD	Department of Defense
DHS	Department of Homeland Security
div	division
DMOC	division medical operations center
DNBI	disease and nonbattle injuries
DOD	Department of Defense
DODD	Department of Defense Directive
DP	decision point
DS	direct support
DSN	defense switched network
DSS	division surgeon section
EAB	echelons above brigade
EAC	echelons above corps
EAD	echelons above division
ECO	environmental compliance officer
EEE	early entry element(s)
EEH	early entry hospital
EEM	early entry module
EMEDS	expeditionary medical support

EMT	emergency medical treatment
EPS	essential personnel services
EPW	enemy prisoner(s) of war
1SG	first sergeant
F	Fahrenheit
FARP	forward arming and refueling point
FAX	facsimile
FEMA	Federal Emergency Management Agency
FHP	force health protection
FM	field manual/frequency modulated
FMC	field medical card
FMI	field manual interim
FRAGO	fragmentary order
FRSS	forward resuscitation
FSB	forward support battalion
FSMC	forward support medical company
FSMT	forward support MEDEVAC team
FST	forward surgical team
G3	Assistant Chief of Staff (Operations and Plans)
GA	ground ambubance
GAMC	ground ambulance medical company
GCC	geographic combatant commander
GRREG	graves registration
GS	general support
GSAB	general support aviation battalion
GWS	Geneva Convention for the Amelioration of the Condition of the Wounded and Sick in Armed Forces in Field
HBCT	heavy brigade combat team
HD	homeland defense
HHD	headquarters and headquarters detachment
HMMWV	high mobility multipurpose wheeled vehicle
HN	host nation
hosp	hospital
HQ	headquarters
HR	human resources
HS	homeland security
HSS	health service support
IAI	installation-level accident investigation
IED	improvised explosive device
IG	Inspector General
IM	information management

IPB	intelligence preparation of the battlefield
IPMC	intratheater patient movement center
ISB	intermediate staging base
IT	information technology
JFC	joint force commander
JHSS	joint health service support
JMD	joint manning document
JMTB	Joint Military Transportation Board
JOA	joint operations area
JP	joint publications
JPMRC	joint patient movement requirements center
JTF	joint task force
LHA	amphibious helicopter assault carrier
LHD	amphibious helicopter assault carrier dock
LIN	line item number
LOC	lines of communication
LOTS	logistics over-the-shore
LPD	amphibious transport dock
LPH	amphibious assault helicopter carrier–aviation
LSD	dock landing ship
LST	tank landing ship
LZ	landing zone
MACA	military assistance to civil authorities
MASCAL	mass casualty
MASF	mobile aeromedical staging facility
MBA	main battle area
MC3T	medical command, control, and communication telemedicine
MC4	medical communications for combat casualty care
MCB	movement control battalion
MCO	major combat operation
MCP	main command post
med	medical
MEDBDE	medical brigade
MEDCOM	medical command
MEDEVAC	medical evacuation
MEDLOG	medical logistics
MEF	Marine expeditionary force
MES	medical equipment sets
METT-TC	mission, enemy, terrain and weather, troops and support available, time available, and civil considerations

MEV	medical evacuation vehicle
MH	mental health
MHS	military health system
MIA	missing in action
MIJI	meaconing, intrusion, jamming, and interference
MMB	multifunctional medical battalion
MOC	medical operations cell
MOPP	mission-oriented protective posture
MOS	military occupational specialty
MP	military police
MRE	meals, ready-to-eat
MRI	medical reengineering initiative
MRO	medical regulating office
MSB	main support battalion
MSC	Military Sealift Command
MTF	medical treatment facility
MTOE	modification table of organization and equipment
MWD	military working dogs
MWR	morale, welfare, and recreation
N/A	not applicable
NATO	North Atlantic Treaty Organization
NAVAIDS	navigational aids
NBC	nuclear, biological, and chemical
NCO	noncommisioned officer
NOE	nap-of-the-earth
NP	neuropsychiatric
NSHS	National Strategy for Homeland Security
NVG	night vision goggles
OCP	operational command post
OEG	operational exposure guidance
OEH	occupational and environmental health
OMF	originating medical facility
OPCON	operational control
OPLAN	operation plan
OPORD	operation order
OPS	operations
OPSEC	operations security
OPTEMPO	operational tempo
OR	operating room
OVE	on vehicle equipment

PASR	personnel accountability and strength reporting
PAD	patient administrator
PCC	precombat checks
PCI	precombat inspections
PDS	personnel daily summary
PERS	personnel
PIM	personnel information management
PIR	priority intelligence requipment
PMCS	preventive maintenance checks and services
PMI	patient movement items
PMM	preventive medicine measures
PMRC	patient movement requestment center
POI	point of injury
POL	petroleum, oils, and lubricants
PR	personnel recovery
prep	preparation
PRM	personnel readiness management
PT	platoon
PSYOP	psychological operations
PZ	pickup zone
PVNTMED	preventive medicine
QSTAG	Quadripartite Standardization Agreement
R5	reception, replacement, return to duty, rest and recuperation, redeployment
R&R	rest and recuperation
RC	reserve component
rep	representative
RPG	rocket propelled grenade
RSTA	reconnaissance and target acquisition
RSOI	reception, staging, onward movement, and integration
RTD	return to duty
S&RO	stability and reconstruction operations
S1	Adjutant (US Army)
S2	Intelligence Officer (US Army)
S3	Operations and Training Officer (US Army)
S4	Supply Officer (US Army)
S6	Information Management (US Army)
SBCT	Stryker Bridage Combat Team
SEC	section
SF	special forces
SIMLM	single integrated medical logistics manager

SINCGARS	Single Channel Ground/Airborne Radio System
SJA	Staff Judge Advocate
SOA	special operations aviation
SOF	special operations forces
SOP	standing operating procedures
SPT	support
SRC	standard requirements code
STANAG	standardization agreement
SURG	surgical
SUST	sustainment
TAA	tactical assembly area
TACC	tanker airlift control center
TAC CP	tactical command post
TAC SOP	tactical standing operating order
TAES	Theater Aeromedical Evacuation System
TC	training circular
tm	team
TMIP	Theater Medical Information Program
tmt	treatment
tng	training
TO	theater operations
TOC	tactical operations center
TOE	tables of organization and equipment
TPFDL	time-phased force deployment list
TPMC	theater patient movement center
TPMRC	theater patient movement requirements center
TSOC	Theater Special Operations Command
TSOP	tactical standing operating procedures
UO	urban operations
US	United States
USAF	United States Air Force
USAMEDDC&S	United States Army Medical Department Center and School
USASC	United States Army Safety Center
USN	United States Navy
USNS	United States Navy Ship
USNORTHCOM	United States Northern Command
USTRANSCOM	United States Transportation Command
vet	veterinary
WARNO	warning order
WIA	wounded in action

WMD	weapons of mass destruction
XO	executive officer
ZULU	universal time

References

SOURCES USED

These are the sources quoted or paraphrased in this publication.

NATO STANAGs

2040. *Stretchers, Bearing Brackets, and Attachment Supports*, 23 September 1982 (Latest Amendment, 6 October 2004)

2087. *Medical Employment of Air Transport in the Forward Area*, 29 June 1983 (Latest Amendment, 10 September 1999)

2128. *Medical and Dental Supply Procedures*, 15 November 1982 (Latest Amendment, 21 November 1991)

2132. *Documentation Relative to Medical Evacuation, Treatment, and Cause of Death of Patients*, 7 August 1974 (Latest Amendment, 15 September 1986)

2454. *Road Movements and Movement Control*, 27 January 2005

2931. *Orders for the Camouflage of the Red Cross and Red Crescent on Land in Tactical Operations*, 18 October 1984 (Latest Amendment, 3 April 1998)

3204. *Aeromedical Evacuation*, 26 September 1973 (Latest Amendment, 10 September 2001)

ABCA QSTAGs

436. *Minimum Labeling Requirements for Medical Materiel* (Amendment 1), 8 October 1980

470. *Documentation Relative to Medical Evacuation, Treatment, and Cause of Death of Patients*, 23 February 1979 (Latest Amendment, 14 August 1989)

PRESIDENTIAL DIRECTIVE

Homeland Security Presidential Directive-5, *Management of Domestic Incidents*, February 28, 2003

DODD

DoD Directive 3025.15, *Military Assistance to Civil Authorities*, February 18, 1997

DoD Directive 5200.31E, *DoD Military Working Dog (MWD) Program*, March 29, 2006

JOINT AND MULTISERVICE PUBLICATIONS

Joint Pub 1-02, *DOD Dictionary of Military and Associated Terms*, 23 March 1994
Joint Pub 4-02, *Doctrine for Health Service Support in Joint Operations*, 26 April 1995
Joint Pub 5-0, *Joint Operation Planning*, 26 December 2006
Joint Pub 5-00.1, *Joint Doctrine for Campaign Planning*, 25 January 2002
Joint Pub 5-00.2, *Joint Task Force (JTF) Planning Guidance and Procedures*, 13 January 1999

ARMY PUBLICATIONS

AR 40-5, *Preventive Medicine*, 22 July 2005

AR 71-32, Force Development and Documentation-Consolidated Policies, 3 March 1997

AR 200-1, *Environmental Protection and Enhancement*, 21 February 1997

AR 385-40, *Accident Reporting and Records*, 1 November 1994

AR 385-64, *U.S. Army Explosives Safety Program*, 1 February 2000

AR 385-95, *Army Aviation Accident Prevention*, 12 December 1999

AR 600-55, *The Army Driver and Operator Standardization Program (Selection, Training, Testing, and Licensing)*, 31 December 1993

DA Pam 385-40, *Army Accident Investigating Reporting*, 1 November 1994

DA Pam 385-40, *Ammunition and Explosives Safety Standards*, 15 December 1999

FM 1-202, *Environmental Flight*, 23 February 1983

FM 1-564, *Shipboard Operations*, 29 June 1007

FM 3-0, *Operations*, 14 June 2001

FM 3-04.301, *Aeromedical Training for Flight Personnel*, 29 September 2000

FM 3-05, *Army Special Operations Forces*, 20 September 2006

FM 3-06.11, *Combined Arms Operations in Urban Terrain*, 28 February 2002

FM 3-07, *Stability Operations and Support Operations*, 20 February 2003

FM 3-11.5, *Multiservice Tactics, Techniques, and Procedures for Chemical, Biological, Radiological, and Nuclear Decontamination*, 4 April 2006

FM 3-34.2, *Combined-Arms Breaching Operations (Including Changes 1—3)*, 31 August 2000

FM 3-97.6, *Mountain Operations*, 28 November 2000

FM 3-100.4, *Environmental Considerations in Military Operations (Including Change 1)*, 15 June 2000

FM 4-01.45, *Multi-Service Tactics, Techniques, and Procedures for Convoy Operations*, 24 March 2005

FM 4-02, *Force Health Protection in a Global Environment*, 13 February 2003

FM 4-02.1, *Combat Health Logistics*, 28 September 2001

FM 4-02.4, *Medical Platoon Leader's Handbook—Tactics, Techniques, and Procedures (Including Change 1)*, 24 August 2001

FM 4-02.6, *The Medical Company—Tactics, Techniques, and Procedures (Including Change 1)*, 4 September 2004

FM 4-02.7, *Health Service Support in a Nuclear, Biological, and Chemical Environment—Tactics, Techniques, and Procedures*, 10 January 2002

FM 4-02.12, *Health Service Support in Corps and Echelons above Corps*, 2 February 2004

FM 4-02.17, *Preventive Medicine Services*, 28 August 2000

FM 4-02.18, *Veterinary Services—Tactics, Techniques, and Procedures*, 30 December 2004

FM 4-02.21, *Division and Brigade Surgeon's Handbook (Digitized)—Tactics, Techniques, and Procedures*, 15 November 2000

FM 4-02.24, *Area Support Medical Battalion—Tactics, Techniques, and Procedures*, 28 August 2000

FM 4-02.25, *Employment of Forward Surgical Teams—Tactics, Techniques, and Procedures*, 28 March 2003

FM 4-25.12, *Unit Field Sanitation Team*, 25 January 2002

FM 4-02.43, *Force Health Protection Support for Army Special Operations Forces*, 28 November 2006

FM 5-0, *Army Planning and Orders Production*, 20 January 2005

FM 8-42, *Combat Health Support in Stability Operations and Support Operations*, 27 October 1997

FM 8-55, *Planning for Health Service Support*, 9 September 1994

FM 8-250, *Preventive Medicine Specialist (Including Change 1)*, 27 January 1986
FM 10-67.1, *Concepts and Equipment of Petroleum Operations*, 2 April 1998
FM 21-10, *Field Hygiene and Sanitation*, 21 June 2000
FM 31-70, *Basic Cold Weather Manual*, 12 April 1968
FM 55-30, *Army Motor Transport Units and Operations (Including Change 1)*, 15 September 1999
FM 90-3, *Desert Operations*, 24 August 1993
FM 90-13, *River Crossing Operations*, 26 January 1998
ST 4-02.121, *Multifunctional Medical Battalion*, October 2005
TC 1-400, *Brigade Aviation Element Handbook*, 27 April 2006
TB MED 507, *Heat Stress Control and Heat Casualty Management*, 7 March 2003
CALL Newsletter 99-9, *Integrating Military Environmental Protection*, 1999
FMI 4-90.1, *Heavy Brigade Combat Team Logistics*, 15 March 2005
FMI 5-0.1, *The Operations Process*, 31 March 2006

DOCUMENTS NEEDED

These documents must be available to the intended users of the publication.

DA Form 285 Series, *U.S. Army Accident Report*
DA Form 1156, *Casualty Feeder Card*
DA Form 2397 Series, *Technical Report of U.S. Army Aircraft Accidents*
DA Form 2404, *Equipment Inspection and Maintenance Worksheet*
DD Form 601, *Patient Evacuation Manifest*
DD Form 1380, *U.S. Field Medical Card*

Index

References are to paragraph numbers unless otherwise stated.

S

T

Field Manual
No. 100- 14

Headquarters
Department of the Army
Washington, DC, 23 April 1998

Risk Management

Contents

DISTRIBUTION RESTRICTION: Approved for public release; distribution is unlimited.

Preface

FM 100-14 applies across the wide range of Army operations. It explains the principles, procedures, and responsibilities to successfully apply the *risk management process* to conserve combat power and resources. The manual applies to both Army and civilian personnel during all Army activities, including joint, multinational, and interagency environments.

The manual is intended to help commanders,[1] their staffs, leaders,[2] and managers develop a framework to make risk management a routine part of planning, preparing, and executing operational missions and everyday tasks. This framework will allow soldiers to operate with maximum initiative, flexibility, and adaptability. Although the manual's prime focus is the operational Army, the principles of risk management apply to all Army activities.

Army operations—especially combat operations—are demanding and complex. They are inherently dangerous, including tough, realistic training. Managing risks related to such operations requires educated judgment and professional competence. The risk management process allows individuals to make informed, conscious decisions to accept risks at acceptable levels.

This manual is not a substitute for thought. Simply reading it will not make one adept in building protection around a mission.[3] Soldiers should compare the doctrine herein against their own experience and think about why, when, and how it applies to their situation and area of responsibility. If the doctrine herein is to be useful, it must become second nature.

The proponent of this manual is HQ TRADOC. Send comments and recommendations on DA Form 2028 directly to Commander, US Army Training and Doctrine Command, ATTN: ATBO-SO, Fort Monroe, VA 23651-5000.

Unless this publication states otherwise, masculine nouns and pronouns do not refer exclusively to men.

---

1 The term *commander* as used herein refers to personnel in a command position.

2 The term *leader* as used herein refers to commanders, personnel in the chain of command (team, squad, section, platoon leader), and staff members having personnel supervisory responsibility.

3 The term *mission* as used herein includes mission, operation, or task.

Introduction

Risk management is not an add-on feature to the decision-making process but rather a fully integrated element of planning and executing operations... Risk management helps us preserve combat power and retain the flexibility for bold and decisive action. Proper risk management is a combat multiplier that we can ill afford to squander.

General Dennis J. Reimer
Chief of Staff, Army
27 July 1995

The Army's fundamental purpose is to fight and win the nation's wars. For this purpose, the country gives the Army critical resources, including those most valuable—its sons and daughters. The Army uses its resources to generate overwhelming combat power to fight and win quickly, decisively, and with minimal losses. The Army's inherent responsibility to the nation is to protect and preserve its resources—a responsibility that resides at all levels. Risk management is an effective process for preserving resources. It is not an event. It is both an art and a science. Soldiers use it to identify tactical and accident risks, which they reduce by avoiding, controlling, or eliminating hazards.

The Army introduced the risk management process into training, the operational environments, and materiel acquisition in the late 1980s. Risk management was originally perceived as solely a safety officer function. However, by the early 1990s, the Army established a goal to integrate risk management into all Army processes and activities and into every individual's behavior, both on and off duty. Since the process was introduced, the personal involvement of commanders in preventing accidents—and their aggressive use of the process—have become driving factors in the steady downward trend in Army accidental losses.

Leaders must understand the importance of the process in conserving combat power and resources. Risk management, like reconnaissance and security, is an ongoing process that continues from mission to mission. Within the mission, leaders must know when the process begins and who has responsibility. It must be integral to the military decision. The process is an important means to enhance situational awareness.

Risk decisions are commanders' business. Such decisions are normally based on the next higher commander's guidance on how much risk he is willing to accept and delegate for the mission. Risk decisions should be made at the lowest possible level, except in extreme circumstances. Training operations, including those at combat training centers (CTCs), may be of such intensity that risk decision are retained at a higher level.

Both leaders and staffs manage risk. Staff members continuously look for hazards associated with their areas of expertise. They then recommend controls to reduce risks. Hazards and the resulting risks may vary as circumstances change and experience is gained. Leaders and individual soldiers become the assessors for ever-changing hazards such as those associated with environment (weather; visibility; contaminated air, water, and soil), equipment readiness, individual and unit experience, and fatigue. Leaders should advise the chain of command on risks and risk reduction measures.

Index

I

K

L

M

N

O

P

R

S

Chapter 1

Risk Management Fundamentals

Sizing up opponents to determine victory, assessing dangers and distances is the proper course of action for military leaders.

Sun Tzu, The Art of War, "Terrain"

Risk management is the process of identifying, assessing, and controlling risks arising from operational factors and making decisions that balance risk costs with mission benefits. Leaders and soldiers at all levels use risk management. It applies to all missions and environments across the wide range of Army operations. Risk management is fundamental in developing confident and competent leaders and units. Proficiency in applying risk management is critical to conserving combat power and resources. Commanders must firmly ground current and future leaders in the critical skills of the five-step risk management process.

Risk is characterized by both the probability and severity of a potential loss that may result from hazards due to the presence of an enemy, an adversary, or some other hazardous condition. Perception of risk varies from person to person. What is risky or dangerous to one person may not be to another. Perception influences leaders' decisions. A publicized event such as a training accident or a relatively minor incident may increase the public's perception of risk for that particular event and time—sometimes to the point of making such risks unacceptable. Failure to effectively manage the risk may make an operation too costly—politically, economically, and in terms of combat power (soldiers lives and equipment). This chapter presents the background, principles, applicability, and constraints relating to the risk management process.

BACKGROUND

Throughout the history of armed conflict, government and military leaders have tried to reckon with the effect of casualties on policy, strategy, and mission accomplishment. Government and military leaders consider battle losses from different perspectives. However, both must balance the following against the value of national objectives:

- Effects of casualties.
- Impact on civilians.
- Damage to the environment.
- Loss of equipment.
- Level of public reaction.

War is inherently complex, dynamic, and fluid. It is characterized by uncertainty, ambiguity, and friction. *Uncertainty* results from unknowns or lack of information. *Ambiguity* is the blurring or fog that makes it difficult to distinguish fact from impression about a situation and the enemy. *Friction* results from change, operational hazards, fatigue, and fears brought on by danger. These characteristics cloud the operating environment; they create risks that affect an army's ability to fight and win. In uncertainty, ambiguity, and friction, both danger and opportunity exist. Hence, a leader's ability to adapt and take risks are key traits. Chapter 2 of FM 100-5 provides information on the challenging circumstances of military operations during conflict.

Historically, the Army has had more accidental losses, including fratricide (friendly fire), than losses from enemy action. See Figure 1-1. These accidental losses are the same types experienced in peacetime

Army	World War II 1942–1945	Korea 1950–1953	Vietnam 1965–1972	Desert Shield/ Storm[1] 1990–1991
Accidents	56%	44%	54%	75%
Friendly Fire	1%	1%	1%	5%
Enemy Action	43%	55%	45%	20%
[1]These numbers include the relatively long buildup time and short period of combat action				

Figure 1-1. Battle and Nonbattle Casualties

during training exercises. These losses are not caused by the enemy or an adversary. Factors include—

- An ever-changing operational environment.
- Effects of a fast-paced, high-operational tempo (OPTEMPO) and a high-personnel tempo (PERSTEMPO) on unit and human performance. Examples include leader or soldier error or failure to train or perform to standards.
- Equipment failure, support failure, and the effects of the physical environment.

PRINCIPLES

The basic principles that provide a framework for implementing the risk management process are—

- *Integrating risk management into mission planning, preparation, and execution.* Leaders and staffs continuously identify hazards and assess both accident and tactical risks. They then develop and coordinate control measures. They determine the level of residual risk for accident hazards in order to evaluate courses of action (COAs). They integrate control measures into staff estimates, operation plans (OPLANs), operation orders (OPORDs), and missions. Commanders assess the areas in which they might take tactical risks. They approve control measures that will reduce risks. Leaders ensure that all soldiers understand and properly execute risk controls. They continuously assess variable hazards and implement risk controls.
- *Making risk decisions at the appropriate level in the chain of command.* The commander should address risk guidance in his commander's guidance. He bases his risk guidance on established Army and other appropriate policies and on his higher commander's direction. He then gives guidance on how much risk he is willing to accept and delegate. Subordinates seek the higher commander's approval to accept risks that might imperil the next higher commander's intent.
- *Accepting no unnecessary risk.* Commanders compare and balance risks against mission expectations and accept risks only if the benefits outweigh the potential costs or losses. Commanders alone decide whether to accept the level of residual risk to accomplish the mission.

APPLICABILITY

Risk management applies to all situations and environments across the wide range of Army operations, activities, and processes. Risk management is useful in developing, fielding, and employing the total Army force. Figure 1-2 summarizes the key aspects of risk management.

DEVELOPMENT

Development concerns include force design, manpower allocation, training and training developments, and combat and materiel developments (equipment and weapons systems) and battle laboratories.

Risk management assists the commander or leader in—

- Conserving lives and resources and avoiding unnecessary risk.
- Making an informed decision to implement a COA.
- Identifying feasible and effective control measures where specific standards do not exist.
- Providing reasonable alternatives for mission accomplishment.

Risk management does not—

- Inhibit the commander's and leader's flexibility and initiative.
- Remove risk altogether, or support a zero defects mindset.
- Require a GO/NO-GO decision.
- Sanction or justify violating the law.
- Remove the necessity for standard drills, tactics, techniques, and procedures.

Figure 1-2. Key Aspects of Risk Management

Force Design

Concerns include risks introduced in trade-off decisions that involve the design and equipping of—

- Tables of organization and equipment (TOE).
- Modification tables of organization and equipment (MTOE).
- Tables of distribution and allowances (TDA) organizations.

Manpower Allocations

Concerns include shortfalls in manning that put unit readiness and full use of combat system capabilities at risk.

Training and Training Developments

Concerns include hazardous and critical training tasks and feasible risk reduction measures that provide leaders with the flexibility to safely conduct tough, realistic training.

Combat and Materiel Developments and Battle Laboratories

Concerns include providing a means to assist in making informed trade-off decisions such as—

- Balancing equipment form, fit, and function.
- Balancing the durability and cost of equipment and spare parts against their reliability, availability, and maintainability requirements.
- Determining the environmental impact.
- Determining whether to accept systems with less than the full capabilities prescribed in requirement documents and experimental procedures.

ARs 70-1 and 385-16 and MIL-STD-882 provide details on risk management application in the Army materiel acquisition process.

FIELDING

Fielding concerns include personnel assignments, sustainment and logistics, training, and base operations.

Personnel Assignments

Concerns include making informed decisions in assigning replacement personnel. For example, a risk is associated with assigning a multiple launch rocket system crewmember as a replacement for a tube artillery cannon crewmember.

Sustainment and Logistics

Concerns include enhancing one's ability to determine support requirements, the order in which they should be received, and the potential impact of logistics decisions on operations.

Training

Concerns include helping leaders determine the—

- Balance between training realism and unnecessary risks in training.
- Impact of training operations on the environment.
- Level of proficiency and experience of soldiers and leaders.

Base Operations

Concerns include prioritizing the execution of base operations functions to get the most benefit from available resources. Examples include allocating resources for pollution prevention, correcting safety and health hazards, and correcting violations of environmental protection regulations. FM 20-400 provides specific guidance on environmental protection in military operations.

EMPLOYMENT

Employment concerns include force protection and deployment, operations, and redeployment.

Force Protection

Concerns include developing a plan that identifies threats and their associated hazards and balancing resource restraints against the risk.

Deployment, Operations, and Redeployment

Concerns include—

- Analyzing the factors of mission, enemy, terrain, troops, and time available (METT-T) to determine both tactical and accident risks and appropriate risk reduction measures.
- Determining the correct units, equipment composition, and sequence.
- Identifying controls essential to safety and environmental protection.

CONSTRAINTS

Risk management does not convey authority to violate the law-of-land warfare or deliberately disobey local, state, national, or host nation laws. It does not justify ignoring regulatory restrictions and applicable standards. Neither does it justify bypassing risk controls required by law, such as life safety and fire protection codes, physical security, transport and disposal of hazardous material and waste, or storage of classified material. Commanders may not use risk management to alter or bypass legislative intent. However, when restrictions imposed by other agencies adversely affect the mission, planners may negotiate a satisfactory COA if the result conforms to the legislative intent.

Risk management assists the commander in complying with regulatory and legal requirements by—

- Identifying applicable legal standards that affect the mission.
- Identifying alternate COAs or alternate standards that meet the intent of the law.
- Ensuring better use of limited resources through establishing priorities to correct known hazardous conditions that will result in projects with the highest return on investment funded first.

Chapter 2

Risk Management Process

First reckon, then risk

Field Marshal Helmuth von Moltke

This chapter provides the essence of the five-step risk management process. It illustrates the application of each step to military operations through the factors of METT-T.

THE FIVE STEPS: AN OVERVIEW

Risk management is the process of identifying and controlling hazards to conserve combat power and resources. The five steps of risk management are—

- Step 1. Identify hazards.
- Step 2. Assess hazards to determine risks.
- Step 3. Develop controls and make risk decisions.
- Step 4. Implement controls.
- Step 5. Supervise and evaluate.

This five-step process is integrated into the military decision-making process as shown in Figure 2-1.

FM 100-40 provides insight into the context in which the risk management process is applied herein. Areas of particular interest in FM 100-40 include—

- Solving tactical problems (Chapter 1).
- The science and art of tactics (Chapter 1).
- Hasty versus deliberate operations (Chapter 1).
- The plan-prepare-execute cycle (Chapter 1).
- Basic tactical control measures (Chapter 2).
- The factors of METT-T (Chapter 2).

Military Decision-Making Process	Risk Management Steps				
	Step 1 Identify Hazards	Step 2 Assess Hazards	Step 3 Develop Controls and Make Risk Decision	Step 4 Implement Controls	Step 5 Supervise and Evaluate
Mission Receipt	X				
Mission Analysis	X	X			
COA Development	X	X	X		
COA Analysis	X	X	X		
COA Comparison			X		
COA Approval			X		
Orders Production				X	
Rehearsal[1]	X	X	X	X	X
Execution and Assessment[1]	X	X	X	X	X

[1]All boxes are marked to emphasize the continued use of the risk management process throughout the mission

Figure 2-1. Risk Management Steps Correlated with Military Decision-Making Tasks

Risk decisions should be based upon awareness rather than mechanical habit. Leaders should act on a keen appreciation for the essential factors that make each situation unique instead of from conditioned response. Throughout the entire operational continuum, the commander must consider US Government civilians and contract support personnel in his risk management process. Hazards can exist, regardless of enemy or adversary actions, in areas with no direct enemy contact and in areas outside the enemy's or adversary's influence. The two types of risk that exist across the wide range of Army operations are *tactical risks* and *accident risks.*

- *Tactical risk* is risk concerned with hazards that exist because of the presence of either the enemy or an adversary. It applies to all levels of war and across the spectrum of operations.
- *Accident risk* includes all operational risk considerations other than tactical risk. It includes risks to the friendly force. It also includes risks posed to civilians by an operation, as well as an operations impact on the environment. It can include activities associated with hazards concerning friendly personnel, civilians, equipment readiness, and environmental conditions.

STEPS 1 AND 2

Steps 1 and 2 together comprise the risk assessment. In Step 1, individuals identify the hazards that may be encountered in executing a mission. In Step 2, they determine the direct impact of each hazard on the operation. The risk assessment provides for enhanced situational awareness. This awareness builds confidence and allows soldiers and units to take timely, efficient, and effective protective measures.

STEPS 3 THROUGH 5

Steps 3 through 5 are the essential follow-through actions to effectively manage risk. In these steps, leaders balance risk against costs—political, economic, environmental, and to combat power—and take appropriate actions to eliminate unnecessary risk. During execution, as well as during planning and preparation, leaders continuously assess the risk to the overall mission and to those involved in the task. Finally, leaders and individuals evaluate the effectiveness of controls and provide lessons learned so that others may benefit from the experience.

THE FIVE STEPS APPLIED

STEP 1. IDENTIFY HAZARDS

A *hazard* is an actual or potential condition where the following can occur due to exposure to the hazard:

- Injury, illness, or death of personnel.
- Damage to or loss of equipment and property.
- Mission degradation.

Hazards are sources of danger or risks due to enemy or adversary presence and other conditions not due to enemy or adversary capabilities. Hazards are found in all operational environments. Combat operations, stability operations, base support operations, and training present unique hazards for units involved in these kinds of missions. Hazards are identified during the first four steps of the military decision-making process: *mission receipt, mission analysis, COA development*, and *COA analysis*.

The ability of unit leaders and staffs to identify hazards is key. One reality of today's missions is that the aspect of a hazard can change rapidly. Things of little risk initially can quickly become major threats due to unforeseen natural or man-made events. Leaders should be aware of this possibility. Complacency to the fact that existing controls may not continue to control hazards in rapidly changing situations should be viewed as a hazard in itself.

The factors of METT-T provide a sound framework for identifying hazards when planning, preparing, and executing operations. When applying risk management to METT-T during mission analysis, leaders and staffs should look for hazards that affect both tactical and accident risks. They must identify all hazards that may present significant risks to the mission.

Mission

Leaders first analyze the assigned mission. They look at the type of mission to be accomplished and consider possible subsequent missions. Certain kinds of operations are inherently more dangerous than others. For example, a deliberate frontal attack, because of the associated movement, is more likely to expose a unit to losses than would a defense from prepared positions. Identifying missions that routinely present great risk is imperative. Leaders also look for hazards associated with complexity of the plan such as—

- A scheme of maneuver that is difficult to understand or too complex for accurate communications down to the lowest level.
- The impact of operating under a fragmentary order (FRAGO).

Enemy

Commanders look for enemy capabilities that pose significant hazards to the operation. For example, "What can the enemy do to defeat my operation?" Common shortfalls that can create hazards during operations against an enemy include failure to—

- Assess potential advantages to the enemy provided by the battlefield environment.
- Fully assess the enemy's capabilities.
- Understand enemy collection capabilities and friendly vulnerabilities to those capabilities.
- Accurately determine the enemy's probable COAs.
- Plan and coordinate active ground and aerial reconnaissance activities.
- Disseminate intelligence about the enemy to lower levels.
- Identifying terrorist threats and capabilities.

Intelligence plays a critical part in identifying hazards associated with tactical risk. Intelligence-preparation-of-the-battlefield (IPB) is a dynamic staff process that continually integrates new information and intelligence that ultimately becomes input to the commander's risk assessment process. Intelligence assists in identifying hazards during operations by—

- Identifying the opportunities and constraints the battlefield environment offers to threat and friendly forces.
- Thoroughly portraying threat capabilities and vulnerabilities.
- Collecting information on populations, governments, and infrastructures.

FMs 34-130 and 34-60, respectively, provide detailed information on IPB and on counterintelligence operations and multidiscipline counterintelligence analysis.

Terrain and Weather

In addition to those due to the enemy or adversaries, the most obvious hazards to military operations are due to terrain and weather. Terrain and weather affect the type of hazard encountered. When the enemy uses terrain to his advantage, the risk is clearly tactical. The aspects of terrain and weather may create situations where accident risks predominate. When looking at this from a purely mission perspective, familiarity of the unit with the terrain and its associated environment must be paramount. Basic issues include—

- How long the unit has operated in the environment and climate.
- Whether the terrain has been crossed before.

Terrain. The five main military aspects of terrain—*observation and fields of fire, cover and concealment, obstacles, key terrain and decisive terrain, and avenues of approach (OCOKA)*—can be used to identify and assess hazards impacting on friendly forces. Chapter 2 of FM 100-40 has details on OCOKA. The terrain analysis includes both map and on-the-ground reconnaissance to identify how well unit capabilities and mission demands can be accommodated by the terrain.

- *Observation and fields of fire.* Hazards associated with this usually involve when the enemy will be able to engage a friendly unit and when friendly unit weapons capabilities allow it to effectively engage the enemy.
- *Cover and concealment.* Hazards associated with cover and concealment are created by the enemy's ability to place direct or indirect fire on friendly forces.
- *Obstacles.* Hazards associated with obstacles may be accident or tactical. They may be due to natural conditions such as rivers or swamps or man-made such as minefields or built-up areas.
- *Key terrain and decisive terrain.* Hazards are a marked advantage terrain provides to the enemy if he controls such terrain or denies its use to friendly forces.
- *Avenues of approach.* Hazards associated with avenues of approach can affect both tactical and accident risks. Such hazards include conditions where an avenue of approach impedes deployment of friendly combat power or where it supports deployment of enemy combat power.

Weather. Weather works hand-in-hand with terrain to create hazards. To identify weather hazards, leaders and soldiers must assess the impact on operating systems. Mistakes include not considering the—

- Adverse effects of heat and cold hazards on the performance of soldiers.
- Effects of climate and weather on maintenance of vehicles and equipment before beginning an operation.
- Hazardous effects of weather on the five military aspects of terrain.

Troops

Leaders analyze the capabilities of available friendly troops. Associated hazards impact both the soldier and unit. Key considerations are level of training, manning levels, the condition and maintenance of vehicles and equipment, morale, availability of supplies and services, and the physical and emotional health of soldiers. Leaders and soldiers must be vigilant to the fact that hazards in these areas can adversely affect a mission, even when all tactical considerations point to success. Mission failure can be caused by—

- *Hazards to the physical and emotional health of soldiers.* Inadequate sanitation facilities, water purification capabilities, medical attention, and evacuation capabilities are key hazards that can arise from incomplete logistical planning. Care of troops requires long-range projections of all classes of supply, with close monitoring of mission changes that could impact availability or depletion of supplies. When beginning an operation immediately upon arriving in theater, hazards include not implementing measures to help soldiers overcome fatigue or acclimatize them to the geographical area and associated climate.
- *Hazards to task organization or units participating in an operation.* Hazards include how long units have worked together under a particular command relationship. During stability operations, task organizations may change often. Hazards include poor communication, unfamiliarity with higher headquarters SOPs, and insufficient combat power to accomplish the mission.
- *Hazards associated with long-term missions.* Long-term missions include nation building, peacekeeping, or insurgency/counterinsurgency operations. Hazards associated with these missions include the turmoil of personnel turnover, lack of continuity of leadership, inexperience, and lack of knowledge of the situation and the unit's operating procedures. An especially insidious hazard is critical-skills atrophy that results from not performing METL-related missions.

Time Available

The hazard is insufficient time to plan, prepare, and execute operations. Planning time is always at a premium. Leaders routinely apply the one-third/two-thirds rule to ensure their subordinate units are given maximum time to plan. Failure to accomplish a mission on time can result in shortages of time for subordinate and adjacent units to accomplish their mission.

Civilians

The commander's legal responsibility is to consider hazards to, and safeguarding of, civilians in his area of operations. *Civilians* include nongovernmental organizations (NGOs), private voluntary organizations (PVOs), US Government civilians, foreign national civilians, the media, and dislocated civilians put at risk by military operations. The commander must consider hazards that can occur across the range of operations, such as—

- *In a wartime environment.* The commander must consider the hazard of collateral damage which may result in creating new adversaries.
- *In a peacetime environment.* The commander must consider the political attitudes and previous actions of civilians in identifying hazards to friendly forces and the populace itself.

Adversaries are hostile elements other than the enemy that may be encountered during any operation. They present additional hazards. They may be organized opposition or individuals that challenge authority. They may include such diverse elements as rioters, criminals, rogues, or gangs that might want to harass a peace enforcement mission.

STEP 2. ASSESS HAZARDS

Step 2 completes the risk assessment. Risk is the chance of hazard or bad consequences. This step examines each hazard in terms of probability and severity to determine the risk level of one or more hazardous incidents that can result from exposure to the hazard. This step is conducted during three steps of the military decision-making process—*mission analysis, COA development,* and *COA analysis.* This step is also conducted after controls are developed.

The incident must be credible in that it must have a reasonable expectation of happening. The end result is an estimate of risk from each hazard and an estimate of the overall risk to the mission caused by hazards that cannot be eliminated. Leaders must also assess the risk to civilians posed by the operation. They may need to assess the operations' impact on the environment. This step is conducted in three substeps.

Substep A

Leaders and staffs assess each hazard in relation to the *probability* of a hazardous incident. The probability levels estimated for each hazard may be based on the mission, COAs being developed and analyzed, or frequency of a similar event. Figure 2-2 provides a summary of the five degrees of probability. The letters in parentheses following each degree (A through E) provide a symbol for depicting probability. For example, the letter *A* represents *frequent* probability.

FREQUENT (A) Occurs very often, continuously experienced	
Single item	Occurs very often in service life. Expected to occur several times over duration of a specific mission or operation. Always occurs.
Fleet or inventory of items	Occurs continuously during a specific mission or operation, or over a service life.
Individual soldier	Occurs very often in career. Expected to occur several times during mission or operation. Always occurs.
All soldiers exposed	Occurs continuously during a specific mission or operation.
LIKELY (B) Occurs several times	
Single item	Occurs several times in service life. Expected to occur during a specific mission or operation.
Fleet or inventory of items	Occurs at a high rate, but experienced intermittently (regular intervals, generally often,).
Individual soldier	Occurs several times in career. Expected to occur during a specific mission or operation.
All soldiers exposed	Occurs at a high rate, but experienced intermittently.
OCCASIONAL (C) Occurs sporadically	
Single item	Occurs some time in service life. May occur about as often as not during a specific mission or operation.
Fleet or inventory of items	Occurs several times in service life.
Individual soldier	Occurs some time in career. May occur during a specific mission or operation, but not often.
All soldiers exposed	Occurs sporadically (irregularly, sparsely, or sometimes).

Figure 2-2. Hazard Probability

SELDOM (D) Remotely possible; could occur at some time	
Single item	Occurs in service life, but only remotely possible. Not expected to occur during a specific mission or operation.
Fleet or inventory of items	Occurs as isolated incidents. Possible to occur some time in service life, but rarely. Usually does not occur.
Individual soldier	Occurs as isolated incident during a career. Remotely possible, but not expected to occur during a specific mission or operation.
All soldiers exposed	Occurs rarely within exposed population as isolated incidents.
UNLIKELY (E) Can assume will not occur, but not impossible	
Single item	Occurrence not impossible, but can assume will almost never occur in service life. Can assume will not occur during a specific mission or operation.
Fleet or inventory of items	Occurs very rarely (almost never or improbable). Incidents may occur over service life.
Individual soldier	Occurrence not impossible, but may assume will not occur in career or during a specific mission or operation.
All soldiers exposed	Occurs very rarely, but not impossible.

Figure 2-2. Hazard Probability (continued)

Substep B

Substep B addresses the *severity* of each hazard. It is expressed in terms of—

- Degree of injury or illness.
- Loss of or damage to equipment or property.
- Environmental damage.
- Other mission-impairing factors such as lost combat power.

The degree of severity estimated for each hazard may be based on knowledge of the results of similar past events. Figure 2-3 provides a summary of the four degrees of hazard severity. The Roman numerals in parentheses following each degree (I through IV) provide a convenient symbol for depicting severity. For example, *I* represents the *catastrophic* degree of severity.

CATASTROPHIC (I)	Loss of ability to accomplish the mission or mission failure. Death or permanent total disability (accident risk). Loss of major or mission-critical system or equipment. Major property (facility) damage. Severe environmental damage. Mission-critical security failure. Unacceptable collateral damage.
CRITICAL (II)	Significantly (severely) degraded mission capability or unit readiness. Permanent partial disability, temporary total disability exceeding 3 months time (accident risk). Extensive (major) damage to equipment or systems. Significant damage to property or the environment. Security failure. Significant collateral damage.
MARGINAL (III)	Degraded mission capability or unit readiness. Minor damage to equipment or systems, property, or the environment. Lost day due to injury or illness not exceeding 3 months (accident risk). Minor damage to property or the environment.
NEGLIGIBLE (IV)	Little or no adverse impact on mission capability. First aid or minor medical treatment (accident risk). Slight equipment or system damage, but fully functional and serviceable. Little or no property or environmental damage.

Figure 2-3. Hazard Severity

Substep C

In this substep leaders and staffs expand what they understand about probable hazardous incidents into estimates of levels of risk for each identified hazard and an estimate of the overall risk for the operation. Estimating risk follows from examining the outcomes of Substeps A and B; that is, both the probability and severity of hazardous incidents. This substep is more art than science. Much depends on the use of historical lessons learned, intuitive analysis, experience, and judgment. Uncertainty can arise in the assessment of both the probability and severity of a hazardous incident. Uncertainty results from unknowns about a situation; from incomplete, inaccurate, undependable, or contradictory information; and from unforeseen circumstances. Therefore, assessment of risk requires good judgment.

Figure 2-4 is a standardized matrix that can be used to assist in this process. Leaders and staffs enter the estimated degree of severity and probability for each hazard in Substeps A and B from the severity row and probability column, respectively. The point where the severity row and probability column intersect defines the level of risk. For example, if the hazard is estimated to have a *critical* severity (II) and a *likely* probability (B), the level of risk is high (H).

Figure 2-5 provides a summary of the levels of risk. It also provides examples of hazardous incidents for each risk level. Several examples illustrate the trade-off between tactical and accident risks.

Risk Assessment Matrix						
		Probability				
Severity		Frequent A	Likely B	Occasional C	Seldom D	Unlikely E
Catastrophic	I	E	E	H	H	M
Critical	II	E	H	H	M	L
Marginal	III	H	M	M	L	L
Negligible	IV	M	L	L	L	L

E – Extremely High Risk
H – High Risk
M – Moderate Risk
L – Low Risk

Figure 2-4. Risk Assessment Matrix

E - Extremely High: Loss of ability to accomplish the mission if hazards occur during mission. A *frequent* or likely probability of catastrophic loss (IA or IB) or *frequent* probability of *critical* loss (IIA) exists.

Example: A commander finds that one of his implied tasks to attack an objective involves crossing a normally shallow riverbed. After looking at the factors of METT-T, he discovers that three days of intense rain have raised the water level to rise above flood stage, with currents far in excess of his ability to safely ford with armored vehicles. After discussing COAs with his staff, he determines the accident risk is extremely high because of the likely probability and catastrophic severity of losing vehicles and killing soldiers. His conclusions are based on his experience with and knowledge of fording armored vehicles under the existing conditions of water depth and current speed.

H - High: Significant degradation of mission capabilities in terms of the required mission standard, inability to accomplish all parts of the mission, or inability to complete the mission to standard if hazards occur during the mission. *Occasional* to *seldom* probability of catastrophic loss (IC or ID) exists. A *likely* to *occasional* probability exists of a critical loss (IIB or IIC) occurring. *Frequent* probability of *marginal* losses (IIIA) exists.

Example: During a preplanned ambush, the leader discovers that the force he intends to ambush has significantly more combat power than his own force can accommodate. He realizes that he could only delay rather than destroy the enemy. He knows his casualty estimates would be very high if the enemy reorganized and counterattacked. He also knows that the size of the enemy force could seriously impact adjacent units conducting a movement to contact. He determines the situation is *high risk* because he estimates (based on his training and experience) there is a likely probability of the enemy reorganizing and counterattacking and the severity of loss to his unit would be critical.

M - Moderate: Expected degraded mission capabilities in terms of the required mission standard will have a reduced mission capability if hazards occur during mission. An *unlikely* probability of catastrophic loss (IE) exists. The probability of a *critical* loss is *seldom* (IID). *Marginal* losses occur with a *likely* or *occasional* probability (IIIB or IIIC). A *frequent* probability of negligible (IVA) losses exists.

Example: A commander in a defensive position receives a warning order to be prepared to counterattack if the enemy attacks again. He chooses to use pre-positioned ammunition caches to support his defense, as opposed to moving his ammunition resupply forward by truck. He determines that the severity of not having an immediate resupply of ammunition available during the counterattack will have a *critical* impact on his combat power. He realizes that if the enemy forces him to abandon his forward positions, the severity of the loss of his

Figure 2-5. Levels of Risk

ammunition caches will critically impact his combat power. He considers that his unit is deployed in excellent defensive positions. He has repelled two attacks that resulted in the destruction of an estimated 50 percent of the enemy's combat power. He receives information that the probability of the enemy attacking is *likely,* but that the probability of the enemy being reinforced and attacking in overwhelming force is remote *(seldom).* The commander concludes that the risk of conducting a counterattack with limited ammunition is greater than the *moderate* risk of the enemy pushing him back.

L - Low: Expected losses have little or no impact on accomplishing the mission. The probability of *critical* loss is *unlikely* (IIE), while that of *marginal* loss is *seldom* (IIID) or *unlikely* (IIIE). The probability of a *negligible* loss is *likely* or *less* (IVB through (IVE).

Example: A mechanized task force (TF) conducting a movement to contact in a desert environment is overtaken by nightfall before reaching its limit of advance (LOA). The terrain along the axis of advance is flat and open. Visibility is about 800 meters under a clear sky illuminated by a full moon. Estimates put the enemy, which has been hastily withdrawing for the past three days, at approximately 30 percent strength. Contact has been light with no defensible terrain along the TF's axis. The TF commander considers all the factors. In addition, the TF is 100 percent operational in using night vision devices. The TF commander estimates that it is *unlikely* that his unit will incur losses of *critical* severity by being surprised by the enemy or lose *critical* combat power due to an accident. He estimates the risk to his force in continuing a nighttime movement is *low.*

Figure 2-5. Levels of Risk (continued)

STEP 3. DEVELOP CONTROLS AND MAKE RISK DECISIONS

> *Risk management is the recognition that decision making occurs under conditions of uncertainty. Decisions must remain consistent with the commander's stated intent and offer a good expectation of success. The risk-taking skill requires competency as a prerequisite.*
>
> FM 100-7, Decisive Force:
> *The Army in Theater Operations,* May 1995

Step 3 is accomplished in two substeps: develop controls and make risk decisions. This is done during the COA development, COA analysis, COA comparison, and COA approval of the military decision-making process.

Substep A - Develop Controls

After assessing each hazard, leaders develop one or more controls that either eliminate the hazard or reduce the risk (probability and/or severity) of a hazardous incident. When developing controls, they consider the reason for the hazard not just the hazard itself.

Types of Controls Controls can take many forms, but fall into three basic categories—*educational controls, physical controls,* and *avoidance.*

- *Educational controls.* These controls are based on the knowledge and skills of the units and individuals. Effective control is implemented through individual and collective training that ensures performance to standard.
- *Physical controls.* These controls may take the form of barriers and guards or signs to warn individuals and units that a hazard exists. Additionally, special controller or oversight personnel responsible for locating specific hazards fall into this category.
- *Avoidance.* These controls are applied when leaders take positive action to prevent contact with an identified hazard.

Criteria for Controls To be effective, each control developed must meet the following criteria:

- *Suitability.* It must remove the hazard or mitigate (reduce) the residual risk to an acceptable level.
- *Feasibility.* The unit must have the capability to implement the control.
- *Acceptability.* The benefit gained by implementing the control must justify the cost in resources and time. The assessment of acceptability is largely subjective. Figure 2-6 gives criteria for determining acceptability of controls for each identified hazard.

Support	Availability of adequate personnel, equipment, supplies, and facilities necessary to implement a suitable controls.
Standards	Guidance and procedures for implementing a control are clear, practical, and specific.
Training	Knowledge and skills are adequate to implement a control.
Leadership	Leaders are competent to implement a control.
Individual	Individual soldiers are sufficiently self-disciplined to implement a control.

Figure 2-6. Criteria for Determining Acceptability of Controls

Examples of Controls Examples of controls include—

- Engineering or designing to eliminate or control hazards.
- Selecting a COA that avoids identified hazards.
- Limiting the number of people and the amount of time they are exposed to hazards, consistent with mission requirements.
- Selecting personnel with appropriate mental, emotional, and physical capabilities.
- Providing protective clothing, equipment, and safety and security devices.
- Providing such services as adequate sanitation facilities and water purification capabilities.
- Providing warning signs and signals.
- Scheduling vehicle and aircraft silhouette drills.
- Planning training, including rehearsals, rock drills, battle drills, and so forth.
- Programming communications links for key civilian organizations.
- Establishing battlefield controls such as areas of operations and boundaries, direct fire control measures, fire support coordination measures, rules of engagement, airspace control measures, bridge classification, traffic control, and so forth.
- Developing terrorist attack warning systems and response plans.

The key is to specify who, what, where, when, and how each control is to be used. For example—

- Planning and scheduling intensive threat and friendly vehicle identification refresher training for all antiarmor and air defense weapons crews before the mission reduces the probability of engaging a friendly vehicle or aircraft (fratricide).
- Programming installation of crashworthy passenger seats in the UH-60 Blackhawk, when mission circumstances do not indicate their removal, can reduce the severity of injuries in crashes.
- Requiring soldiers to wear flak vests and helmets during movement to contact, or when riding in vehicles in areas where enemy fire is likely, can reduce the probability and severity of a wound from small arms fire or fragments.

- Establishing strong continuity documents and planning overlap tours for key leaders facilitate smooth transitions during extended operations.

Residual RiskOnce the responsible leader develops and accepts controls, he determines the residual risk associated with each hazard and the overall residual risk for the mission.

- *Residual risk* is the risk remaining after controls have been selected for the hazard. Residual risk is valid (true) only if the controls for it are implemented. As controls for hazards are identified and selected, the hazards are reassessed as in Step 2 and the level of risk is then revised. This process is repeated until the level of residual risk is acceptable to the commander or leader or cannot be further reduced. See Figures A-3 through A-5.
- *Overall residual risk* of a mission must be determined when more than one hazard is identified. The residual risk for each of these hazards may have a different level, depending on the assessed probability and severity of the hazardous incident. Overall residual mission risk should be determined based on the incident having the greatest residual risk. Determining overall mission risk by averaging the risks of all hazards is not valid. If one hazard has high risk, the overall residual risk of the mission is high, no matter how many moderate or low risk hazards are present.

Substep B - Make Risk Decision

A key element of the risk decision is determining if the risk is justified. The commander must compare and balance the risk against mission expectations. He alone decides if controls are sufficient and acceptable and whether to accept the resulting residual risk. If he determines the risk level is too high, he directs the development of additional controls or alternate controls, or he modifies, changes, or rejects the COA.

Leaders can use the risk assessment matrix in Figure 2-4—in conjunction with their commanders' guidance—to communicate how much risk they are willing to delegate. For example, a commander may place constraints on his subordinates that restrict their freedom of action to accept risk in instances where the risk might imperil his intent, his higher commander's intent, or a critical capability of the unit.

STEP 4. IMPLEMENT CONTROLS

Leaders and staffs ensure that controls are integrated into SOPs, written and verbal orders, mission briefings, and staff estimates. The critical check for this step, with oversight, is to ensure that controls are converted into clear, simple execution orders understood at all levels. Implementing controls includes coordination and communication with—

- Appropriate superior, adjacent, and subordinate units and those executing the mission.
- Logistics Civil Augmentation Program (LOGCAP) organizations and civilian agencies that are part of the force.

The media, NGOs, and PVOs must be included in coordination when their presence impacts or is impacted by the force.

Leaders must explain how supervisors will implement controls. Examples of control implementation include—

- Conducting vehicle and aircraft silhouette drills.
- Conducting rehearsals, rock drills, battle drills, and so forth.
- Conducting intensive threat and friendly vehicle identification refresher training for all antiarmor and air defense weapons crews.
- Conducting orientation for replacement personnel.
- Installing and maintaining communications links for key civilian organizations.
- Operating in convoys of four vehicles minimum.
- Carrying weapons and wearing flak jackets and helmets when outside secure compounds.

STEP 5. SUPERVISE AND EVALUATE

> *Leaders must supervise the execution of their orders. The more untrained the troops, the more detailed this supervision must be.*
>
> *Infantry in Battle*, 1939

During mission preparation and execution, leaders must ensure that their subordinates understand how to execute risk controls. Leaders continuously assess risks during the conduct of operations, especially during long-term missions. Leaders maintain situational awareness. They guard against complacency to ensure that risk control standards are not relaxed or violated. To gain insight into areas needing improvement, leaders must continuously evaluate their units' effectiveness in managing mission risks.

Supervise

Leaders supervise mission rehearsal and execution to ensure standards and controls are enforced. Techniques may include spot-checks, inspections, situation reports and brief-backs, buddy checks, and close supervision. During the mission, leaders continuously monitor controls to ensure they remain effective. They modify them as necessary. Leaders and individuals anticipate, identify, and assess new hazards to implement controls. They continually assess variable hazards such as fatigue, equipment serviceability, and the environment. Leaders modify controls to keep risks at an acceptable level.

During sustained operations, leaders continue planning to ensure that controls emplaced at the beginning of the mission apply to changes in the operation's current situation and to hazardous conditions. Leaders must maintain an extraordinary degree of discipline. They must avoid complacency, which can result from boredom and overconfidence. Leaders must ensure that soldiers do not relax their vigilance due to performing repetitive tasks—despite changing roles and missions, unit turbulence and turnover, and declining skills. Leaders maintain a close overwatch on controls put in place to reduce risks over a prolonged period. For example, during stability operations, land mine hazards may not be solved in the near term, but may require continual attention. Other examples of long-term hazards that may be encountered include—

- Climatic extremes.
- NBC and hazardous waste contamination.
- Diseases native to a particular area of operation or indigenous population.
- Terrorist threats.

Evaluate

After a mission, leaders and individuals evaluate how well the risk management process was executed. They—

- Determine how to ensure that successes are continued to the next mission.

- Capture and disseminate lessons learned so that others may benefit from the experience.
- Consider the effectiveness of the risk assessment in identifying and accurately assessing the probability and severity of hazards that resulted in mission degradation.
- Determine whether the level of residual risk of each hazard and of the overall mission were accurately estimated.
- Evaluate the effectiveness of each control in reducing or removing risk, including whether controls were effectively communicated, implemented and enforced.

Leaders and individuals determine why some controls were ineffective and what should be done when the hazard is encountered again. A control may be altered; the way it is implemented or supervised may be changed to make it effective; or a completely different control may be more effective. Leaders must energize the system to fix systemic problems that hinder combat effectiveness.

Figure 2-7 shows that the risk management process continues throughout a mission as well as from mission to mission. It is integral to the military decision-making process. Its application requires good judgment and intuitive analysis borne of confidence, experience, and situational awareness.

TOOLS AND PITFALLS

The appendix provides examples of risk management tools to help leaders assess identified hazards, develop controls, and make risk decisions. The tools should be tailored to suit particular situations and missions. The examples in Figures A-3 through A-5 are tools to manage risk at the tactical level. The example in Figure A-6 is a tool to manage risk at the operational level. Units may develop additional tools suitable for their needs.

Units train to a standard. They operate and train regardless of the degree of real or perceived difficulty. Risk reduction begins with commanders identifying their METLs. Commanders use the risk management process to assess the degree of risk related to each METL their unit must perform. From this assessment, risk reducing standard operating procedures evolve.

Figure A-7 provides an example of risk management considerations integrated into a mission training plan (MTP) task.

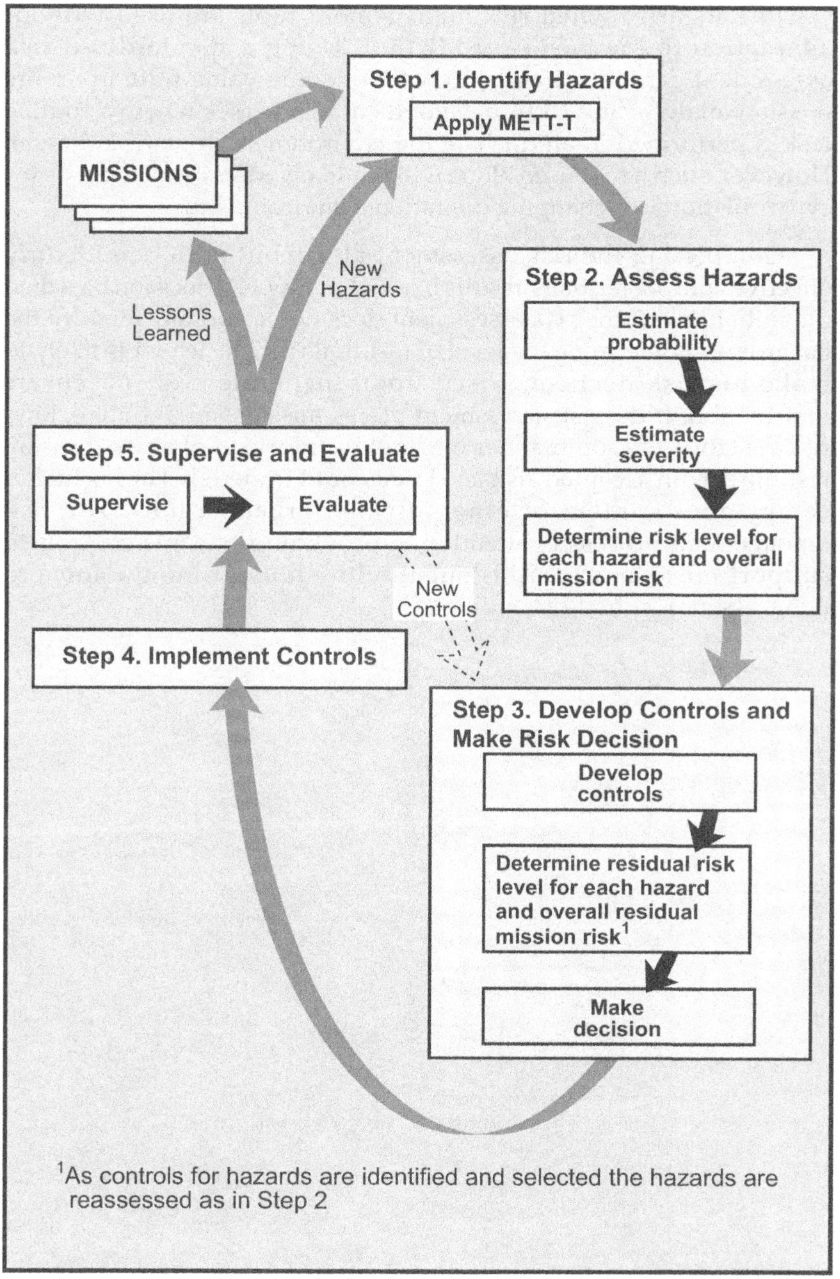

Figure 2-7. Continuous Application of Risk Management

Pitfalls arise when risk management tools are used without adaptation to the factors of METT-T. Using a standardized risk assessment card or checklist may be of some value initially in the mission analysis and COA development or in cases where a routine task is performed in an unchanging environment or static situation. However, such a tool used alone will not likely identify all hazards for every mission in a changing operational environment.

Completing the risk assessment alone, but failing to identify effective controls, usually results in a GO or NO-GO decision based on the initial risk. If the risk assessment does not accurately identify the hazards and determine the level of residual risk, the leader is likely to make his risk decision based upon incomplete or inaccurate information. If the risk assessment places missions in a routine, low-risk category, the commander may not be informed of a risk decision resulting in an accepted risk level that could imperil his or his higher commander's intent or other affected organizations. The risk management process is intended to provide reasonable controls to support mission accomplishment without exposing the force to unnecessary residual risk.

Chapter 3

Risk Management Implementation

It is imperative to develop twenty-first century leaders who trust their subordinates' abilities and judgment. Such leaders must be willing to underwrite their subordinates' honest errors and coach them on to excellence, without tolerating incompetence or laxity. We must recognize that Army leaders are not perfect, and that activity at the ragged edge of audacity sometimes leaves debris in its wake.

GEN William W. Hartzog
Commanding General
US Army Training and Doctrine Command

This chapter presents the moral and ethical implications of risk management. It outlines essential responsibilities and considerations necessary to effectively integrate and assess risk management within the Army. Leaders should tailor these essentials to develop specific how-to procedures suited to their circumstances, available resources, and mission.

MORAL AND ETHICAL IMPLICATIONS FOR LEADERS

To be successful, risk management must be underwritten by the chain of command. Leaders should not expect that all missions will be accomplished with zero defects—free from errors, flaws, or less-than-perfect performance. Demanding such rigid standards leads to oversupervision and paralysis; it produces timid leaders, afraid to make tough decisions in crisis and unwilling to take risks necessary for success in military operations. A zero defects mindset creates conditions that will lead inevitably, in the larger sense, to failure in battle and higher casualties. Leaders are morally bound to support a subordinate's decision to accept risks that are within his commander's intent and guidance, as he understands it.

Leaders accept that things may go wrong, even with the certain knowledge that a subordinate has done all within his power to prevent an incident. In such an event, the leader steps forward and accepts the responsibility along with his subordinate. Furthermore, risk management does not justify taking actions to facilitate an unethical or immoral action. FM 22-100 addresses the moral and ethical aspects of protecting the force within the overall framework of how to apply leadership at all levels to meet mission requirements.

> *Everyday as we respond to the nation's needs, we expose our soldiers to hazards in uncertain and complex environments. We do this with the full knowledge that there are inherent risks associated with any military operation. The nature of our profession will not allow for either complacency or a cavalier acceptance of risk.*
>
> General Dennis J. Reimer
> Chief of Staff, Army

RESPONSIBILITIES

One critical task for all operations is minimizing risk. Every military plan must make this a priority. It is an inherent part of every mission and a basic responsibility of commanders. Whether for training or operational deployments, commanders issue clear risk guidance. Minimizing risk—eliminating unnecessary risk—is the responsibility of everyone in the chain of command. This responsibility runs from the highest commander, through his subordinate leaders, to the soldier.

The commander and his staff must look at tactical risks and accident risks. They use the same risk management process to manage both types. Commanders—with the assistance of their leaders and staffs—manage accident risks. Commanders determine how and where they are willing to take tactical risks.

Leaders and soldiers at all levels are responsible and accountable for managing risks by ensuring that hazards and associated risks are—

- Identified during planning, preparation, and execution of operations.
- Controlled during preparation and execution of operations.

Soldiers are responsible for executing risk controls to standards. They must continuously assess variable hazards such as fatigue, equipment serviceability, and the environment. They must take care of one another and make risk decisions consistent with the higher commander's guidance.

Sometimes commanders are not properly advised in situations where the assumption of risk may affect or imperil their units, the intent of their higher commander, or the operations of an adjacent unit. This is most often attributed to—

- The *risk denial syndrome* in which leaders do not want to know of the risk.
- A staff member who believes that the risk decision is part of his job and does not want to bother his commander or section leader.
- A subordinate who does not fully understand the higher commander's guidance.
- Complacency—outright failure to recognize a hazard or the level of risk involved, or overconfidence in one's abilities or the unit's capabilities to avoid or recover from a hazardous incident.
- Use of a standardized risk assessment tool, such as a risk assessment card, that is not tailored to the unit's mission or adapted to the factors of METT-T and which may put missions in the routine low-risk category.

COMMANDERS

The commander directs the organization and sets priorities and the command climate (values, attitudes, and beliefs). Successful preservation of combat power requires embedding risk management into unit behavior. This requires commitment and creative leadership—innovative planning, careful management. It also requires the chain of command's demonstrated support of the risk management process. Only then will the Army begin to capture the full power of risk management. Commanders establish a command climate favorable for risk management integration by—

- Demonstrating consistent and sustained risk management behavior through leading by example—habitually doing risk management—and actively participating throughout the risk management process.
- Providing clear guidance, when appropriate, on where or what risk to accept.

- Obtaining and providing to subordinates the necessary assets to control risk.
- Knowing their own limitations, their leaders' and soldiers' limitations, and their unit's capabilities.
- Preventing a zero-defects mindset from creeping into their command's culture.
- Allowing subordinates to make mistakes and learn from them.
- Demonstrating full confidence in subordinates' mastery of their trade and their ability to execute a chosen COA.
- Keeping subordinates informed; consulting with subordinate leaders before making a decision, if feasible.
- Listening to subordinates.

Commanders are responsible and accountable for their own actions and those of units under their charge. Commanders must weigh the repercussions of casualties, damage to the environment, and loss of equipment. They must also consider the level of public reaction to loss against national, strategic, operational, or tactical objectives. Commanders are also responsible for keeping soldiers from falling into complacency. During sustained operations in particular, complacency can creep in as a result of accepting the status quo. For example, a unit can be deployed for several months and nothing may happen. The unit then gets very satisfied with itself and its performance. It gets conditioned to its initial circumstances—being well-armed, well-equipped, well-disciplined, and well-trained. It thinks it has risks under control and does not need to change. Such complacency, and the associated loss of situational awareness, can result in leaders and soldiers taking a gamble instead of a prudent risk. A risk is the accepted result of an informed decision; a gamble is an uninformed bet or guess on a hopeful outcome. Leaders and soldiers must clearly understand the difference.

Command is often exercised in conditions of uncertainty and ambiguity, where violence, danger, fear, and friction abound, and under the ever present time constraints driven by OPTEMPO. Risk decisions are frequently required by and dependent on the immediate situation. Judgment is required; a formula, rule, or checklist, by itself, is not appropriate under such circumstances.

Avoiding the zero-risk mindset requires the exercise of positive leadership. The commander's approach to managing risk should be through empowering leaders by pushing risk decisions as far down the chain of command as feasible within the next higher commander's guidance. Commanders must insist that subordinates exercise their freedom of action to act decisively and aggressively to complete assigned missions and promote success of the larger force. Subordinates must consider hazards outside their assigned responsibilities that impact the mission. The result is to encourage coordination and communication—laterally and up and down the chain of command. This requires and encourages initiative, which demands well-trained, determined, disciplined soldiers.

Risk management is a two-way street. It is important that those involved in mission preparation and execution be fully aware of the amount of command involvement and actions necessary to control or remove hazards. The higher commander's guidance specifies the degree of damage or risk to subordinate units that he is willing to accept during the current operation. Subordinates ensure they understand and implement their commander's intent and guidance. If, during the planning process, the accident and/or tactical residual risk exceeds that which the higher commander is willing to accept, the subordinate informs his commander. He requests the resources necessary to mitigate the risk. If, during mission execution, the subordinate determines the risk is too great, he directs the development of additional or alternate controls or modifies or changes the COA. He should notify his next higher commander of his decision. Requiring subordinates to report to the higher commander when a risk decision point is reached during mission execution can result in paralysis.

The objective of managing risk is not to remove all risk, but to eliminate unnecessary risk. Commanders conduct tough, realistic training, knowing that they may put lives and property at risk in the course of military operations. Nothing is worth the cost of a life as the result of taking unnecessary risk. If an action will result in an unacceptable risk, measures should be taken to mitigate it. If the risk cannot be mitigated to an acceptable level, the action should not be executed. Circumstances may occur during mission execution when a decision to stop and defer execution of the operation should be made to avoid taking unwarranted risk. Such a situation will generally occur at the tactical level. For example, circumstances may determine if a trade-off between maintaining the momentum of the attack or risking fratricide or serious accidents is justified. For example, during the deployment of TF Eagle to Bosnia-Herzegovina, there was pressure to complete the Sava River bridge as quickly as possible. The commander assessed the situation and decided to call for a short rest period because his soldiers were tired and getting careless. Consequently, they completed the bridge ahead of schedule under the most difficult conditions imaginable and without injuries.

LEADERS

Many of the tasks identified for commanders apply to all leaders. Leaders' specific responsibilities in managing risk include—

- Establishing clear, feasible risk management policies and goals.
- Conducting detailed planning within time constraints; assessing each mission and task in terms of its risk; continuously reassessing risk as the mission and conditions change and experience is gained.
- Making informed risk decisions and establishing and clearly communicating risk guidance.
- Training the risk management process. Ensuring subordinates understand the who, what, where, when, how, and why of managing risk and how the process applies to their circumstances and assigned responsibilities.
- Examining how subordinates manage risk and how soldiers protect themselves.
- Supervising and evaluating the unit's execution of risk controls during the mission to correct areas needing improvement.
- Advising his chain of command on risks and risk-reduction measures.
- Providing subordinates with feedback on their performance and ways to improve.
- Assessing the effectiveness of their unit's risk management program.
- Capturing and disseminating lessons learned to ensure they are continued from mission to mission so that others may benefit from the experience.

STAFFS

The chief of staff or executive officer is responsible for supervising integration of risk management across the staff. He coordinates development of risk reduction controls with emphasis on deconflicting controls that affect multiple functional areas and adjacent units. The staff officer helps the commander eliminate unnecessary risks by—

- Analyzing his functional area and applying risk management during the military decision-making process.
- Identifying constraints in the higher commander's risk guidance.
- Including hazards and their risks in the mission analysis briefing.
- Including a risk assessment for the commander's estimate.
- Considering the risk assessment in the operations estimate.
- Including risks and recommending ways to reduce their impact in the staff estimate.
- Implementing risk controls by coordinating and integrating them into the appropriate paragraphs and graphics of the OPORD and into products such as SOPs and OPLANs.
- Establishing procedures and standards that are clear and practical.
- Determining the effectiveness of hazard/risk controls and continuously assessing their suitability, feasibility, and acceptability.
- Supervising, evaluating, and assessing the integration of risk management during an operation.
- Continuously identifying hazards, assessing initial and residual risks for each hazard, recommending control measures to reduce the risk to the force.
- Identifying and assessing hazards associated with complacency, especially during extended operations, and recommending appropriate actions to the commander.

At the operational level, staffs focus on hazards and their risks across the spectrum of protecting the force

INDIVIDUALS

The level of expertise and maturity of the individual influence his proficiency in managing risk. Managing risk is subjective because its basis is individual judgment. Young soldiers are routinely charged with executing hazard controls and risk reduction measures. By nature, some them are impulsive risk takers. Their limited experience, coupled with an infallible attitude, can significantly increase the level of risk they are willing to accept. Their sense of indestructibility, motivation *(esprit de corps),* and willingness to achieve the mission at any cost also play a part.

Some soldiers and individuals ignore existing standards and improvise their own. Due to inexperience or complacency, they become susceptible to—

- Overestimating their ability to respond to or recover from a hazardous incident—they become overconfident.
- Underestimating the level of risk posed by a hazard.

It is imperative that individuals understand and execute controls implemented by leaders and staffs.

Individuals must maintain situational awareness and self-discipline when they perform their duties. They must—

- Understand and apply risk management.
- Execute controls directed by their leaders, that is, *perform to standards.*
- Carry risk management over into training and activities—both on and off duty.
- Look out for others—anyone has authority to halt something that is inherently unsafe.

INTEGRATION INTO TRAINING AND OPERATIONS

ARFOR commanders/leaders must continuously employ risk management approaches to effectively preclude unacceptable risks to personnel and property, including protecting forces preparing for or en route to combat.

FM 100-7, *Decisive Force: The Army in Theater Operations,* May 1995

Integrating risk management into training and operations—

- Preserves the lives and well-being of everyone.
- Conserves equipment, facilities, environmental resources, and combat power.

Risk management must not be treated as an afterthought. It must be planned for up front. Leaders and managers of materiel acquisition, base operations, and industrial operations must budget risk control costs up front at the level of expected payback over the duration of the activity, or the life cycle of materiel/weapons system.

When integrating risk management into sustained operations, leaders must consider increases in turbulence, personnel turnover, critical skill atrophy, and mission development. Leaders must continuously assess—

- The complexity of mission development and associated changing interrelationships with other agencies.
- The inclusion of civilian contractors, for example, LOGCAP, as part of the force.
- The presence of the media, NGOs, and PVOs.

These diverse elements need to be integrated into the risk management process.

Two key considerations relevant to managing risk in complex operational environments include—

- Understanding the culture of the indigenous population or society and its way of doing business. Leaders should respect their way of life and not interfere with local customs. Such interference could risk damage to relationships and increase the potential for introducing instability into the local society. Leaders must not, however, intentionally allow these considerations to endanger their force or its mission.
- The dynamics of managing risk and the way the leader and his subordinates manage risk. Leaders manage risk by—
  - Having the right combination of well-trained, disciplined, well-armed, and well-equipped forces.
  - Issuing clear guidance to minimize risk.
  - Determining and implementing risk controls for carrying out the mission.

— Maintaining situational awareness, especially of vulnerabilities.

— Avoiding unnecessary accident risk.

Risk management works best when incorporated into existing command training and operational planning cycles. They then act as risk management tools and provide for—

- Development of policy, goals, objectives, and priorities in the commander's quarterly training guidance.
- The commander's training assessment, by identifying hazards and risk controls.
- Systematic observation and assessment of the unit's risk management performance and feedback into the training management cycle and SOPs.

Leaders and soldiers must have the skills, knowledge, and attitude to effectively manage risks inherent in all operations. Effective training helps soldiers become proficient. It qualifies them technically and tactically, and as leaders, to accomplish the mission without unnecessary risk.

Doctrine developers must integrate risk management into planning for all Army processes, especially the training management cycle defined in FMs 25-100 and 25-101. Unit leaders and their staffs must continually assess and evaluate the integration of risk management into short-, near-, and long-term training plans. They must continually review METLs to ensure that training is supported by realistic risk management objectives. In the past, unprepared or improperly trained units paid a high price for veteran status. The Army must learn from past experiences if it is to avoid repeating such losses.

ASSESSMENT OF THE RISK MANAGEMENT PROCESS

To assess the risk management process is to determine a unit's current level of proficiency in implementing the process. The term *assessment,* as discussed here, differs from *evaluation* as used in Step 5 of the process. Evaluation is used to measure demonstrated ability to accomplish specified objectives within a discrete operation or exercise. Assessment, as used here, also differs from the aspect of assessing hazards for probability and severity, as addressed in Step 2 of the process.

How well risk is managed affects readiness. Leaders need to know the current status and effectiveness of their organization's risk management program. They self-assess their unit's effectiveness in managing risk in order to gain insight into areas for improvement and get feedback on subordinates' understanding and application of risk guidance. The assessment objectives are to determine how—

- Effectively risk management is embedded into planning and preparing for operations.
- Well risk management is understood by subordinate leaders and soldiers.
- Effectively risk management is used to execute operations.

Leaders assess the effectiveness of their units by reviewing how well hazards are identified and risk controls are—

- Specified in oral and written OPORDs, OPLANs, and SOPs.
- Communicated to lowest level of chain of command.
- Included in short-, near-, and long-term training plans.
- Implemented into training and activities on and off duty.
- Embedded into protect-the-force programs such as safety and health and antiterrorism.
- Part of after-action reviews and fed into lessons learned.

Risk management cannot be seen as a competitive program whereby a unit or leader is judged or compared in a competitive sense. Focus is strictly on both reduction of risk and risk behavior.

Take calculated risks. That is quite different from being rash.

General George S. Patton, Jr.

Appendix

Examples of Risk Management Application

The examples in this appendix are designed to help those charged with managing risk.

TRACKING TOOL

The work sheet instructions are in Figure A-1.

Work Sheet Instructions	
Block	
A – D	Self explanatory
E	Identify task relating to the mission or task in Block A
F	**Identify Hazards** – Identify hazards by reviewing METT-T factors for the mission or task. Additional factors include historical lessons learned, experience, judgment, equipment characteristics and warnings, and environmental considerations.
G	**Assess Hazards** – Assessment includes historical lessons learned, intuitive analyses, experience, judgment, equipment characteristics and warnings, and environmental considerations. Determine initial risk for each hazard by applying risk assessment matrix (Figure 2-4). Enter the risk level for each hazard.
H	**Develop Controls** – Develop one or more controls for each hazard that will either eliminate the hazard or reduce the risk (probability and/or severity) of a hazardous incident. Specify who, what, where, why, when, and how for each control. Enter controls.
I	**Determine Residual Risk** – Determine the residual risk for each hazard by applying the risk assessment matrix (Figure 2-4). Enter the residual risk level for each hazard.
J	**Implement Controls** – Decide how each control will be put into effect or communicated to the personnel who will make it happen (written or verbal instruction; tactical, safety, garrison SOPs, rehearsals). Enter controls.
K	**Determine Overall Mission/Task Risk** – Select the highest residual risk level and circle it. This becomes the overall mission or task risk level. The commander decides whether the controls are sufficient to accept the level of residual risk. If the risk is too great to continue the mission or task, the commander directs development of additional controls or modifies, changes, or rejects the COA.
	Supervise and Evaluate – This last step is not on the worksheet. Plan how each control will be monitored for implementation (continuous supervision, spot-checks) and reassess hazards as the situation changes. Determine if the controls worked and if they can be improved. Pass on lessons learned.

Figure A-1. Risk Management Work Sheet Instructions

The work sheet (Figure A-2) provides a starting point to logically track the process of hazards and risks. It can be used to document risk management steps taken during planning, preparation, and execution of training and combat missions and tasks.

A. Mission or Task:			**B. Date/Time Group** Begin: End:		**C. Date Prepared:**
D. Prepared By: (Rank, Last Name, Duty Position)					
E. Task	**F. Identify Hazards**	**G. Assess Hazards**	**H. Develop Controls**	**I. Determine Residual Risk**	**J. Implement Controls** ("How To")
K. Determine overall mission/task risk level after controls are implemented (circle one)					
LOW (L)	MODERATE (M)		HIGH (H)		EXTREMELY HIGH (E)

SAMPLE

Figure A-2. Sample Risk Management Work Sheet

Examples provided in Figures A-3 through A-6 should help individuals manage risk at the tactical level.

A. Mission or Task: Prepare defensive positions		**B. Date/Time Group** **Begin:** 010035R May XX **End:** 010600R May XX		**C. Date Prepared:** 29 April XX	
D. Prepared By: (Rank, Last Name, Duty Position) LT Jones, Plt Ldr					
E. Task	**F. Identify Hazards**	**G. Assess Hazards**	**H. Develop Controls**	**I. Determine Residual Risk**	**J. Implement Controls** ("How To")
Construct nonstandard antivehicular wire obstacle	Back injuries and wire cuts during materiel offload	Moderate (M)	Use proper lift and carry methods and wear concertina wire gloves and safety goggles	Low (L)	Unit TACSOP, ARTEP 5-145 DRILL (pg 2-44) ARTEP 5-335-11-MTP
	Blunt trauma and cuts in pounding of U-shaped pickets	Moderate (M)	Wear helmet and increase situational awareness	Low (L)	Unit TACSOP, ARTEP 5-145 DRILL (pg 2-44) ARTEP 5-335-11-MTP
	Cuts when unrolling concertina	Moderate (M)	Wear concertina wire gloves and maintain situational awareness	Low (L)	Unit TACSOP, ARTEP 5-145 DRILL (pg 2-44) ARTEP 5-335-11-MTP
	Cuts when installing concertina	Moderate (M)	Wear concertina wire gloves and maintain situational awareness	Low (L)	Unit TACSOP, ARTEP 5-145 DRILL (pg 2-44) ARTEP 5-335-11-MTP
	Cuts when installing barbed wire	Moderate (M)	Wear concertina wire gloves and maintain situational awareness	Low (L)	Unit TACSOP, ARTEP 5-145 DRILL (pg 2-44) ARTEP 5-335-11-MTP
K. Determine overall mission/task risk level after controls are implemented (circle one) LOW (L) (circled) MODERATE (M) HIGH (H) EXTREMELY HIGH (E)					

Figure A-3. Example of Completed Risk Management Work Sheet for Squad/Platoon

A. Mission or Task:		B. Date/Time Group		C. Date Prepared:	
Conduct a deliberate attack		**Begin:** 010035R May XX **End:** 010600R May XX		29 April XX	
D. Prepared By: (Rank, Last Name, Duty Position) CPT William Wallace, Cdr					
E. Task	**F. Identify Hazards**	**G. Assess Hazards**	**H. Develop Controls**	**I. Determine Residual Risk**	**J. Implement Controls** ("How To")
Conduct obstacle breaching operations	Obstacles	High (H)	Develop and use obstacle reduction plan	Low (L)	Unit TACSOP, OPORD, training handbook
	Inexperienced soldiers	High (H)	Additional instruction and increased supervision	Moderate (M)	Modified training schedule, additional instruction
	Operating under limited visibility	Moderate (M)	Use NVDs; use IR markers on vehicles	Low (L)	Unit TACSOP, OPORD
	Steep cliffs	High (H)	Rehearse use of climbing ropes	Moderate (M)	FM 90-6, *Mountain Operations* TC 90-6-1, *Mountaineering*
	Insufficient planning time	High (H)	Plan and prepare concurrently	Moderate (M)	OPORD, troop-leading procedures
K. Determine overall mission/task risk level after controls are implemented (circle one)					
LOW (L)	MODERATE (M) (circled)	HIGH (H)	EXTREMELY HIGH (E)		

Figure A-4. Example of Completed Risk Management Work Sheet for Company/Team

A. Mission or Task: Redeploy unit to home station			B. Date/Time Group Begin: 010035R May XX End: 010600R May XX	C. Date Prepared: 29 April XX	
D. Prepared By: (Rank, Last Name, Duty Position) MAJ Woolsey, S3					
E. Task	**F. Identify Hazards**	**G. Assess Hazards**	**H. Develop Controls**	**I. Determine Residual Risk**	**J. Implement Controls** ("How To")
Conduct convoy operations-movement from Grafenwohr to home base	Fatigue leading to poor judgment/ accident-causing errors	Moderate (M)	1. Ensure all drivers receive adequate rest the night before movement 2. Brief planned rest stops and actions to take if driver becomes too tired/ill or has doubts about alertness 3. Have NCOs check physical alertness of drivers prior to movement	Low (L)	Unit SOP, OPORD on Tactical Road March, STP 21-2-MQS (Tactical Road March)
	Fast moving traffic mix of cars and large trucks with trailers	Moderate (M)	1. Use extreme caution when passing 2. Continue to scan (be aware of) traffic 3. Brief all drivers and vehicle commanders on lane-changing procedures	Low (L)	German laws and unit SOP
	Accidents and breakdowns blocking road	Moderate (M)	1. Brief drivers on actions to take a. Pull as far off the road as possible b. Get out on passenger side c. Post triangle sign at least 100 meters behind vehicle d. Stay as far away from vehicle and traffic as possible e. Brief all personnel on maintenance plan/action if vehicle is disabled f. Brief all personnel on procedures for vehicle exits on the traffic side	Low (L)	Unit SOP, OPORD on Tactical Road March

Figure A-5. Example of Completed Risk Management Work Sheet for Battalion/Task Force

A. Mission or Task: Redeploy unit to home station (continued)		B. Date/Time Group Begin: 010035R Dec XX End: 010600R Dec XX		C. Date Prepared: 29 Nov XX	
D. Prepared By: (Rank, Last Name, Duty Position) MAJ Woolsey, S3					
E. Task	F. Identify Hazards	G. Assess Hazards	H. Develop Controls	I. Determine Residual Risk	J. Implement Controls ("How To")
Conduct convoy operations-movement from Grafenwohr to home base	Accidents and breakdowns blocking road (continued)	Moderate (M)	2. Leaders ensure that proper maintenance are done prior to move (PMCS), dispatching and licensing procedure. 3. Contact team follows convoy and repairs or removes disabled vehicles from highway.	Low (L)	
	Weather conditions– high winds, sleet, snow	High (H)	1. Reduce speed according to environmental conditions less stopping distance available. 2. Road conditions especially right after rains – hydroplaning – review procedures for skids or hydroplanes. 3. Maintain convoy interval to allow for braking and assist civilian traffic in passing.	Moderate (M)	Unit SOP and vehicle operations manual
	High winds– reduction visibility, sudden reduction due to passing traffic	Moderate (M)	1. Change lanes carefully; ensure clearance and avoid erratic moves 2. Be aware of winds and effects on vehicles. 3. Drive defensively.	Low (L)	German traffic laws
	20% downgrade at checkpoint 1	Moderate (M)	1. Use lower gears. 2. Reduce speed to 25 mph	Low (L)	Unit SOP and vehicle operations manual

Figure A-5. Example of Completed Risk Management Work Sheet for Battalion/Task Force (continued)

A. Mission or Task: Redeploy unit to home station (continued)		B. Date/Time Group Begin: 010035R Dec XX End: 010600R Dec XX		C. Date Prepared: 29 Nov XX	
D. Prepared By: (Rank, Last Name, Duty Position) MAJ Woolsey, S3					
E. Task	**F. Identify Hazards**	**G. Assess Hazards**	**H. Develop Controls**	**I. Determine Residual Risk**	**J. Implement Controls** ("How To")
Conduct convoy operations- movement from Grafenwohr to home base	Heavy traffic, traffic jams, and congestion	Moderate (M)	1. Use extreme caution when passing 2. Use headlights at all times 3. Brief convoy speeds and vehicle intervals 4. Brief actions to take if vehicle pulls in front a. Maintain clearance from vehicle b. Maintain braking distance 5. Stay in right lane in construction zones 6. Ensure AMVs 2 or more meters wide remain in right lane 7. Continuously scan surroundings (situational awareness)	Low (L)	Provide risk assessment to convoy leaders. Provide safety brief information to leaders and drivers
K. Determine overall mission/task risk level after controls are implemented (circle one) LOW (L) MODERATE (M) [circled] HIGH (H) EXTREMELY HIGH (E)					

Figure A-5. Example of Completed Risk Management Work Sheet for Battalion/Task Force (continued)

A. Mission or Task: Peace Enforcement	B. Date/Time Group Begin: 010035R May XX End:	C. Date Prepared: 29 April XX
D. Prepared By: (Rank, Last Name, Duty Position) G-3 Div/Corps/EAC		

E. Task	F. Identify Hazards	G. Assess Hazards	H. Develop Controls	I. Determine Residual Risk	J. Implement Controls ("How To")
Occupy area of operations	Assault on facilities	Moderate (M)	Identify and isolate combatants, terrorists, supporters Harden potential terrorist targets	Low (L)	Coordination with local law enforcement agencies FM 5-103, *Survivability General /Order*
	Ambush	Moderate (M)	Kevlar helmets and flak jackets required outside cantonment area Minimum of 4 vehicles for movement Two qualified drivers per vehicle	Low (L)	Rehearsals Immediate action drills General orders
	Mines	High (H)	Countermine plan Mine awareness training Predeployment training	Moderate (M)	FM 20-32, *Mine/Countermine Operations* Lessons learned

Figure A-6. Example of Completed Risk Management Work Sheet for Division/Corps/EAC Task Force

A. Mission or Task: Peace Enforcement			B. Date/Time Group Begin: 010035R May XX End:		C. Date Prepared: 29 April XX
D. Prepared By: (Rank, Last Name, Duty Position) G-3 Div/Corps/EAC					
E. Task	**F. Identify Hazards**	**G. Assess Hazards**	**H. Develop Controls**	**I. Determine Residual Risk**	**J. Implement Controls ("How To")**
Occupy area of operations	Fratricide	Moderate (M)	Identify friend/foe Antifratricide training LNO augmentations	Low (L)	Lessons learned ROE Weapons control procedures AOR orientation
	Seasonal risk (cold weather injuries)	Moderate (M)	Cold weather clothing and equipment Soldier/leader awareness training	Low (L)	Severe weather alert/action plan TB Med 507
	Seasonal risk (hot weather injuries)	Moderate (M)	Hot weather clothing and equipment Soldier/leader awareness training	Low (L)	Severe weather alert/action plan TB Med 507
	Disease	Moderate (M)	Immunizations Field sanitation Preventive medicine	Low (L)	Spring thaw plan Focused medical training Vector control program

Figure A-6. Example of Completed Risk Management Work Sheet for Division/Corps/EAC Task Force (continued)

A. Mission or Task: Peace Enforcement		**B. Date/Time Group** **Begin:** 010035R May XX **End:**		**C. Date Prepared:** 29 April XX	
D. Prepared By: (Rank, Last Name, Duty Position) G-3 Div/Corps/EAC					
E. Task	**F. Identify Hazards**	**G. Assess Hazards**	**H. Develop Controls**	**I. Determine Residual Risk**	**J. Implement Controls** ("How To")
Occupy area of operations	Vehicle accidents	Moderate (M)	Driver qualifications Driver fatigue reduction Convoy safety procedures Monitor route conditions Vehicle inspections	Low (L)	Driver qualification training Driver sleep plans Two qualified drivers per vehicle Convoy safety briefings Route reconnaissance Vehicle inspections
K. Determine overall mission/task risk level after controls are implemented (circle one) LOW (L) MODERATE (M) (circled) HIGH (H) EXTREMELY HIGH (E)					

Figure A-6. Example of Completed Risk Management Work Sheet for Division/Corps/EAC Task Force (continued)

MISSION TRAINING PLAN

Figure A-7 shows risk management integrated into an MTP task. The example portrays a possible method for integrating the risk management process and products into MTPs. The text in bold italics indicates suggested areas in which to integrate risk management.

TASK: C^2 THE BATTALION (7-1-1901)(FM 7-20)

ITERATION: 1 2 3 4 5 (circle)

TRAINING STATUS: T P U (circle)

CONDITION: The brigade issues an OPORD or FRAGO

TASK STANDARD :

a. The battalion plan accomplishes the directed mission and specified tasks according to the brigade commander's concept and intent. The plan is received and understood by the leadership of the battalion, who makes the plan successful. It is coordinated with higher, adjacent, and supporting elements.

b. The plan is as fully prepared as time allows to initiate the mission at the directed time.

c. The battalion controls and synchronizes subordinate and supporting elements so that it accomplishes the mission and preserves the force. ***Include commander's risk guidance.***

d. The battalion keeps higher, adjacent, subordinate, supporting, and supported headquarters informed of essential information key to controlling the battle or making required decisions.

SUBTASKS AND STANDARDS: GO NO-GO

+1. Battalion leaders issue the warning order.

a. A complete warning order is issued within 15 minutes of receipt of the brigade order.

b. Warning order is received by all platoons within 45 minutes of issuance of battalion warning order.

+ Critical task.

* Leader task.

Figure A-7. Integration of Risk Management into MTP Task

*2. Battalion commander analyzes mission and gives initial guidance.

a. Guidance includes restated mission, which includes brigade commander's intent for battalion and identifies all specified and implied tasks.

b. Guidance is given within 30 minutes of receipt of order.

c. Guidance includes instructions on information requirements and initially required preparation actions (movement, resupply) to start. ***Guidance includes chain of command authorized to accept risk (extremely high, high, moderate, and low) affecting higher command's intent, resources, or adjacent units' missions.***

+3. Battalion accomplishes reconnaissance and other actions to gather needed information.

a. Reconnaissance actions begin to physically gain information on the enemy and terrain as early as possible.

b. Commander conducts a personal reconnaissance, when possible. If not, the commander conducts a detailed map reconnaissance.

c. Subordinate leaders perform a personal reconnaissance when possible. See subordinate company (ARTEP 7-10-MTP) and platoon (ARTEP 7-8-MTP) T&EOs.

d. Staff coordinates with subordinate, higher, supporting, and adjacent headquarters to gather information for planning. ***Staff analyzes mission to identify hazards.***

e. Staff provides operations, intelligence, and CSS estimates to include all critical METT-T factors and ***risk assessment considerations (identified hazards/risks).***

+4. Battalion commander develops and wargames courses of action and selects one.

Figure A-7. Integration of Risk Management into MTP Task (continued)

a. Tactically feasible courses of action (include CS and maneuver) are made and wargamed with the available staff (commander, S3, and FSO are best for quick planning sequences; XO, S2, engineer, S4, S3 (Safety), and ADA officer are best in more deliberate situations). ***Each course of action contains assessment of hazards, risk level, and control measures identified to lower or control the risk.***

b. Best COA is selected.

c. COA is wargamed and refined by the command and staff. The staff must understand the concept to produce a sound OPORD and rehearse. ***The staff continually assesses the situation to identify new hazards, assess the risk from each hazard, develop acceptable control measures, and reassess for residual risk to the mission. COA should pose minimum risk to soldiers, equipment, and mission accomplishment. Risk acceptance decisions for the COA are made at the appropriate level in the chain of command.***

*5. Staff develops an OPLAN and OPORD from the commander's guidance.

OPLAN and OPORD successfully accomplish the mission according to higher commander's intent. ***Hazards and risk control measures are included in the appropriate paragraphs and graphics.***

*6. Battalion commander and staff issue implementing FRAGOs (accident risk controls).

*7. Battalion commander issues the OPORD/FRAGO.

a. OPORD/FRAGO is issued IAW the one-third, two-thirds rule and makes full use of daylight time.

b. OPORD/FRAGO accomplishes all directed missions and tasks, complies with the brigade commander's intent, controls risks, and is doctrinally sound. It is based on evaluator judgment and on comparison of brigade and battalion OPORDs.

Figure A-7. Integration of Risk Management into MTP Task (continued)

c. All subordinate and supporting elements receive the OPORD/FRAGO.

d. OPORD/FRAGO contains task organization, mission, concept, ***accident risk controls,*** intent for maneuver, supporting fires, and obstacles; missions and tasks for each subordinate; fire support and CSS instructions; and coordinating instructions to synchronize the efforts of maneuver forces and CS.

e. If more time is available, the battalion commander issues a fully developed OPORD (although an initial FRAGO may be issued to allow subordinates to begin preparation and followed by a full OPORD.)

f. Order is given at a location that reduces travel time, allows observation of the zone/sector, and promotes OPSEC. (Depending upon the METT-T factors, observation of the zone/sector may not be possible.)

g. Battalion commander should perform brief-backs and war gaming, ***to include identifying hazards and control measures,*** immediately after the order to improve subordinate understanding and reaction.

h. Subordinate leaders and staff should perform lateral coordination before leaving the orders site.

+8. Commander and staff coordinate and refine the plan.

a. Time is well used to continue gathering information and to improve the plan (contingency plans, ***hazard identification and controls,*** fire plans) .

b. New information is disseminated and coordinated with higher, adjacent, and supporting headquarters to include—

- Changes or refinements in plan.
- Information on the enemy in the sector or zone.

Figure A-7. Integration of Risk Management into MTP Task (continued)

- Information that impacts on planning and execution ***(risk acceptance decisions/hazard controls)*** of subordinate elements.
- Adjustments/changes in the plans.

+*9. Battalion executes changes in task organization.

a. Main CP coordinates link-up location, time, and responsible element.

b. Attachments/new elements are received at the coordinated location and time; updated on current situation, OPORDs, and SOIs; and resupplied.

c. Detachments reach the link-up point at the time and place directed.

+*10. Battalion performs and commander and staff perform, supervise, and monitor preparations.

a. Command group/XO performs brief-backs with subordinate commanders, leaders, and key staff.

b. Main CP maintains status of preparations.

c. Elements make full use of time to prepare for the operation. Subjective judgment of the evaluator is based on the analysis of preparation charts and available preparation time.

+*11. Battalion sees the battlefield.

a. Command group is positioned to see and move.

b. Companies and other subordinates accurately report critical information on actions and changes in combat status within five minutes. See subordinate MTPs.

c. Main CP collects, analyzes, and passes processed critical information.

d. Subordinates execute intelligence collection plan. See

Figure A-7. Integration of Risk Management into MTP Task (continued)

subordinate element MTPs. ***Subordinates integrate the risk management process when developing plans and executing tasks.***

+*12. Battalion leaders command and control the execution.

a. Subordinate elements report enemy and friendly actions, change in status, and any other factor that would require change within three minutes.

b. Battalion leaders win the battle by directing the maneuver of units, controlling direct and indirect fires, ***properly integrating risk management into planning, preparation, and execution,*** and directing other CS actions to cope with new METT-T/risk factors. Indicators are:

(1) Elements not following OPLAN and OPORD are corrected.

(2) Responses to new METT-T hazards are directed soon after the new situation occurs.

(3) No friendly casualties inflicted by friendly direct or indirect fires or other accidents.

(4) Number/percentage of direct fire weapons engaging the enemy.

(5) Number/percentage of indirect fire rounds fired and percentage hitting/suppressing the enemy.

(6) Number of enemy casualties.

(7) Number of friendly casualties.

c. The C^2 and CSS assets are controlled to support maneuver effort. Indicators are—

(1) Effective CSS and C^2.

(2) C^2 or CSS elements not destroyed by enemy direct fires.

d. FRAGOs are clear, concise, ***include risk management,*** and are quickly executed by subordinates.

Figure A-7. Integration of Risk Management into MTP Task (continued)

e. Changes that affect the battle are disseminated within five minutes.

+13. Subordinate commanders, leaders, and staff laterally coordinate actions during the battle.

All battle actions requiring coordination between elements are coordinated.

+*14. Battalion coordinates with adjacent and supporting headquarters.

All battle actions requiring coordination with other headquarters are laterally and promptly coordinated.

+*15. Battalion reports.

Battalion CPs submit all critical and required reports to brigade. They report events to adjacent and supporting elements that impact on them in time for those units to react. *They advise the chain of command, as appropriate, on risks and risk-reduction measures.*

Figure A-7. Integration of Risk Management into MTP Task (continued)

Glossary

AAR	after-action review
ADA	air defense artillery
ARFOR	army forces headquarters
ARTEP	Army Training and Evaluation Program
assessment	an analytical process to determine an organization's current levels of proficiency on a specific objective (for example, a training objective or risk management implementation) (CJCSM 3500.03)
base operations support	the provision of administrative and logistical services; includes supply operations, maintenance of materiel, personnel support, base services and administrative services rendered by or through activities of the supporting installation
BASOPS	base operations support
C^2	command and control
COA	course of action
CofS	chief of staff
combat power	the total means of destructive and/or disruptive force that a military unit or formation can apply against an opponent at a given time; a combination of the effects of maneuver, firepower, protection, and leadership
controls	actions taken to eliminate hazards or reduce their risk
CP	command post
CSS	combat service support
CTC	combat training center

danger	exposure or vulnerability to harm or risk; the balance between the chance or probability of a hazardous incident and the result of the hazardous incident
EAC	echelons above corps
evaluation	the process used to measure the demonstrated ability to accomplish specified objectives such as training within a discrete event or exercise (CJCSM 3500.03); measurement of the demonstrated ability of soldiers or units to perform a task and supporting skill and knowledge or learning objective against the established standard
exposure	the frequency and length of time personnel and equipment are subjected to a hazard
FM	field manual
FRAGO	fragmentary order
fratricide	the employment of friendly weapons and munitions with the intent to kill the enemy or destroy his equipment or facilities, which results in unforeseen and unintentional death or injury to friendly personnel
friction	the accumulation of chance errors, unexpected difficulties, enemy actions, and confusion of battle
FSO	fire support officer
G3	general staff operations section
hazard	any actual or potential condition that can cause injury, illness, or death of personnel, damage to or loss of equipment, property or mission degradation (FM 101-5); a condition or activity with potential to cause damage, loss or mission degradation (Joint Pub 1-02)
inherently dangerous	an activity or task containing a danger to life or limb that is a permanent and inseparable element of the activity
IPB	intelligence-preparation-of-the-battlefield
IR	infrared

LOA	limit of advance
LOGCAP	Logistics Civil Augmentation Program
METL	mission-essential task list
METT-T	mission, enemy, terrain, troops, and time available
MOS	military occupational specialty
MTP	mission training plan
NBC	nuclear, biological, chemical
NVD	night vision device
OCOKA	O - observation and fields of fire, C - cover and concealment, O - obstacles, K - key terrain and decisive terrain, A - avenues of approach
OPCON	operational control
operational tempo	the pace of an operation or operations; OPTEMPO includes all of the activities the unit is conducting; OPTEMPO can be a single activity or a series of operations
OPLAN	operations plan
OPORD	operations order
OPTEMPO	operational tempo
personnel tempo	unit work load level and number of deployed days per year
PERSTEMPO	personnel tempo
PIR	priority intelligence requirements
probability	the likelihood that a hazardous incident will occur
PVO	private voluntary organization
residual risk	the level of risk remaining after controls have been identified and selected for hazards that may result in loss of combat power
risk	chance of hazard or bad consequences; the probability of exposure to chance of injury or loss from a hazard; risk level is expressed in terms of hazard probability and severity (FM 101-5)

risk assessment identification and assessment of hazards (first two steps of risk management process); an identified hazard is assessed to determine the risk (both the probability of occurrence and resulting severity) of a hazardous incident due to the presence of the hazard

risk decision the decision to accept or not accept the risks associated with an action; made by the commander, leader, or individual responsible for performing that action

risk management the process of identifying, assessing, and controlling risks arising from operational factors and making decisions that balance risk cost with mission benefits

risk management integration the embedding of risk management principles and practices into Army operations, culture, organizations, systems, and individual behavior

S3 operations officer

severity the expected consequence of an event (hazardous incident) in terms of degree of injury, property damage, or other mission-impairing factors (loss of combat power and so on) that could occur

situational awareness ability to have accurate and real-time information on friendly, enemy, neutral, and noncombatant locations; a common, relevant picture of the battlefield scaled to specific level of interest and special need

SOP standing operating procedure

T&EO test and evaluation outline

TACSOP tactical standing operating procedure

TC training circular

TF task force

US United States

XO executive officer

References

SOURCES USED

FM 20-400. *Military Environmental Protection.* 1997 (draft).

FM 22-9. *Soldier Performance in Continuous Operations.* 12 December 1991.

FM 22-100. *Army Leadership.* 1997 (draft).

FM 25-100. *Training the Force.* 15 November 1988.

FM 25-101. *Battle-Focused Training.* 30 September 1990.

FM 34-60. *Counterintelligence.* 5 February 1990.

FM 34-130. *Intelligence Preparation of the Battlefield.* 23 May 1989.

FM 100-5. *Operations.* August 1997 (draft).

FM 100-40. *Tactics.* 29 October 1997 (draft).

FM 100-7. *Decisive Force: The Army in Theater Operations.* 31 May 1995.

FM 101-5. *Staff Organization and Operations.* 31 May 1997.

READINGS RECOMMENDED

Army Regulation 70-1. *Systems Acquisition Policy and Procedures.* 1997.

Army Regulation 385-16. *System Safety Engineering and Management.* 3 May 1990.

CSA Statement on Risk Management. 27 July 1995.

HQDA Letter 5-97-1. Risk Management Integration Responsibilities. 1 May 1997.

MIL-STD-882C. *System Safety Program Requirements.* 19 January 1993.

"Risk Management for Brigades and Battalions." *Center for Army Lessons Learned (CALL) Newsletter 95-9.* June 1995.

Made in the USA
Monee, IL
05 November 2026

67808335R00144